Monitoring Progress Toward Successful K-12 STEM Education

A NATION ADVANCING?

Committee on the Evaluation Framework for Successful K-12 STEM Education

Board on Science Education and Board on Testing and Assessment

Division of Behavioral and Social Sciences and Education

NATIONAL RESEARCH COUNCIL
OF THE NATIONAL ACADEMIES

THE NATIONAL ACADEMIES PRESS
Washington, D.C.
www.nap.edu

THE NATIONAL ACADEMIES PRESS • 500 Fifth Street, NW • Washington, DC 20001

NOTICE: The project that is the subject of this report was approved by the Governing Board of the National Research Council, whose members are drawn from the councils of the National Academy of Sciences, the National Academy of Engineering, and the Institute of Medicine. The members of the committee responsible for the report were chosen for their special competences and with regard for appropriate balance.

This study was supported by Grant Nos. DRL-1233221 and DRL-1247500 between the National Academy of Sciences and the National Science Foundation. Any opinions, findings, conclusions, or recommendations expressed in this publication are those of the author(s) and do not necessarily reflect the views of the organizations or agencies that provided support for the project.

International Standard Book Number-13: 978-0-309-26481-5
International Standard Book Number-10: 0-309-26481-2

Additional copies of this report are available from the National Academies Press, 500 Fifth Street, NW, Keck 360, Washington, DC 20001; (800) 624-6242 or (202) 334-3313; http://www.nap.edu.

Printed in the United States of America

Suggested citation: National Research Council. (2013). *Monitoring progress toward successful K-12 STEM education: A nation advancing?* Committee on the Evaluation Framework for Successful K-12 STEM Education. Board on Science Education and Board on Testing and Assessment, Division of Behavioral and Social Sciences and Education. Washington, DC: The National Academies Press.

THE NATIONAL ACADEMIES

Advisers to the Nation on Science, Engineering, and Medicine

The **National Academy of Sciences** is a private, nonprofit, self-perpetuating society of distinguished scholars engaged in scientific and engineering research, dedicated to the furtherance of science and technology and to their use for the general welfare. Upon the authority of the charter granted to it by the Congress in 1863, the Academy has a mandate that requires it to advise the federal government on scientific and technical matters. Dr. Ralph J. Cicerone is president of the National Academy of Sciences.

The **National Academy of Engineering** was established in 1964, under the charter of the National Academy of Sciences, as a parallel organization of outstanding engineers. It is autonomous in its administration and in the selection of its members, sharing with the National Academy of Sciences the responsibility for advising the federal government. The National Academy of Engineering also sponsors engineering programs aimed at meeting national needs, encourages education and research, and recognizes the superior achievements of engineers. Dr. Charles M. Vest is president of the National Academy of Engineering.

The **Institute of Medicine** was established in 1970 by the National Academy of Sciences to secure the services of eminent members of appropriate professions in the examination of policy matters pertaining to the health of the public. The Institute acts under the responsibility given to the National Academy of Sciences by its congressional charter to be an adviser to the federal government and, upon its own initiative, to identify issues of medical care, research, and education. Dr. Harvey V. Fineberg is president of the Institute of Medicine.

The **National Research Council** was organized by the National Academy of Sciences in 1916 to associate the broad community of science and technology with the Academy's purposes of furthering knowledge and advising the federal government. Functioning in accordance with general policies determined by the Academy, the Council has become the principal operating agency of both the National Academy of Sciences and the National Academy of Engineering in providing services to the government, the public, and the scientific and engineering communities. The Council is administered jointly by both Academies and the Institute of Medicine. Dr. Ralph J. Cicerone and Dr. Charles M. Vest are chair and vice chair, respectively, of the National Research Council.

www.national-academies.org

COMMITTEE ON THE EVALUATION FRAMEWORK FOR SUCCESSFUL K-12 STEM EDUCATION

ADAM GAMORAN (*Chair*), Department of Sociology and Wisconsin Center for Education Research, University of Wisconsin–Madison

RENA DORPH, The Research Group, The Lawrence Hall of Science; University of California, Berkeley

MARK DYNARSKI, Pemberton Research, LLC, East Windsor, New Jersey

DAVID FRANCIS, Department of Psychology, University of Houston

SHARON LEWIS, Council of the Great City Schools, Detroit, Michigan

BARBARA MEANS, Center for Technology in Learning, SRI International, Menlo Park, California

MEREDITH PHILLIPS, School of Public Affairs, University of California, Los Angeles

WILLIAM SCHMIDT, Departments of Statistics and Education, and the Education Policy Center, Michigan State University

THOMAS SMITH, Peabody College of Education and Human Development, Vanderbilt University

RUTH LÓPEZ TURLEY, Department of Sociology, Rice University

SUZANNE WILSON, Department of Teacher Education, Michigan State University

NATALIE NIELSEN, *Study Director*

STUART W. ELLIOTT, *Senior Staff Officer*

MARTIN STORKSDIECK, *Senior Staff Officer*

REBECCA KRONE, *Program Associate*

CONTENTS

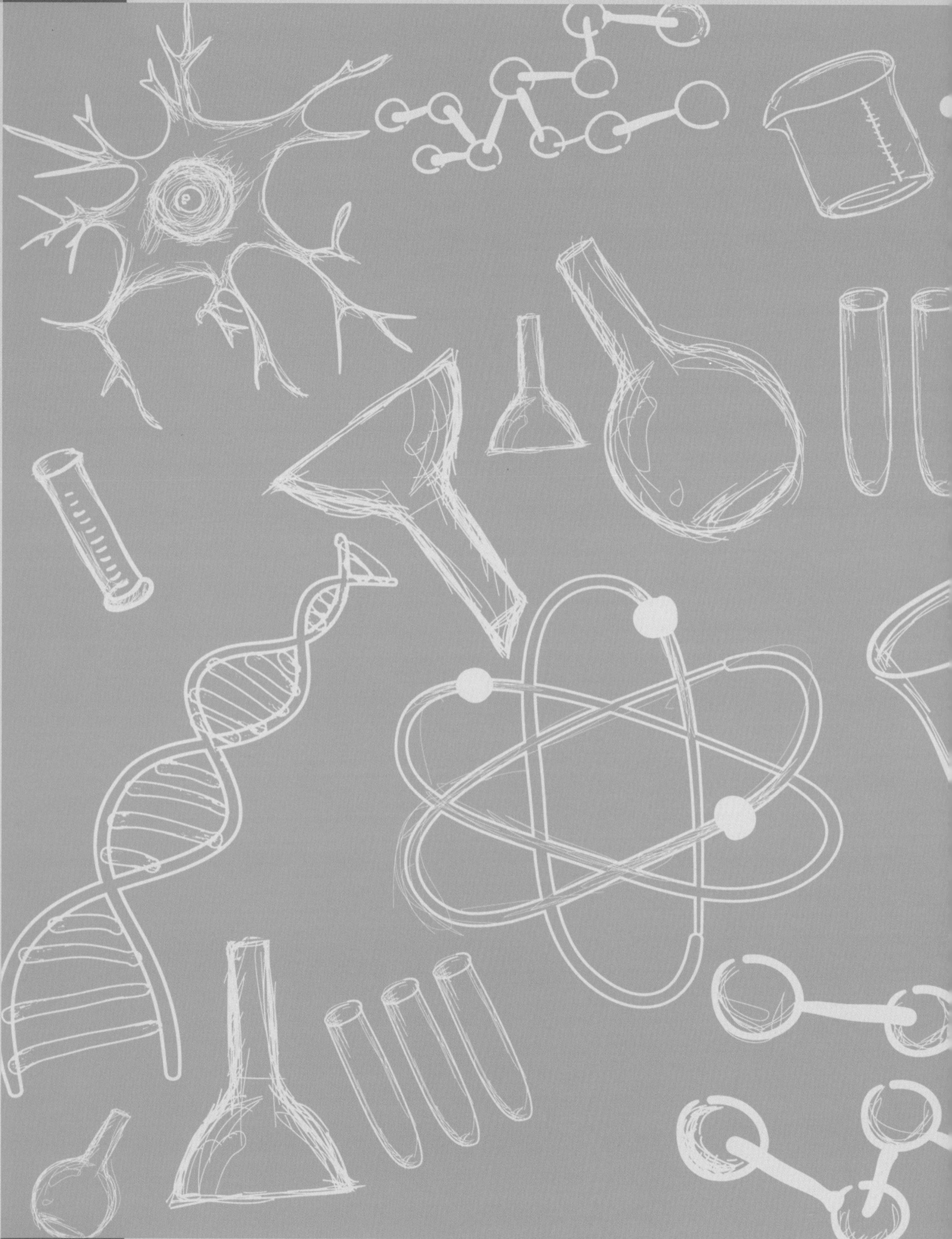

EXECUTIVE SUMMARY

Following a 2011 report by the National Research Council (NRC) on successful K-12 education in science, technology, engineering, and mathematics (STEM), Congress asked the National Science Foundation to identify methods for tracking progress toward the report's recommendations. In response, the NRC convened the Committee on the Evaluation Framework for Successful K-12 STEM Education to take on this assignment.

The committee developed 14 indicators linked to the 2011 report's recommendations, shown in the table on page 2. By providing a focused set of key indicators related to students' access to quality learning, educators' capacity, and policy and funding initiatives in STEM, the committee addresses the need for research and data that can be used to monitor progress in the K-12 STEM education system and make informed decisions about improving it.

Our recommended indicators provide a framework for Congress and relevant federal agencies to create and implement a national-level monitoring and reporting system with the capability to:

- assess progress toward key improvements recommended in the 2011 National Research Council report *Successful K-12 STEM Education*;
- measure student knowledge, interest, and participation in the STEM disciplines and STEM-related activities;
- track financial, human capital, and material investments in K-12 STEM education at the federal, state, and local levels;
- provide information about the capabilities of the STEM-education workforce, including teachers and principals; and
- facilitate strategic planning for federal investments in STEM education and workforce development, when used with labor force projections.

All 14 indicators are intended to form the core of this system. However, the indicators highlighted in bold in the table—2, 4, 5, 6, 9, and 14—reflect the committee's highest priorities. With the exception of Indicator 14, the priority indicators are nearest to the core of student learning. As such, they represent the points of greatest leverage to improve the education system and student outcomes in the STEM disciplines, and to make progress toward the goals of increasing the number of underrepresented students who pursue science and engineering degrees and careers, expanding the STEM-capable workforce, and increasing science literacy. The committee deemed Indicator 14 as a high priority because it assesses progress in filling critical gaps in knowledge about programs and practices that contribute to the goals of STEM education.

Recommendations from *Successful K-12 STEM Education* (2011)	**Indicators**
Districts Should Consider All Three Models of STEM-Focused Schools	1. Number of, and enrollment in, different types of STEM schools and programs in each district.
Districts Should Devote Adequate Instructional Time and Resources to Science in Grades K-5	2. **Time allocated to teach science in grades K-5.**
	3. Science-related learning opportunities in elementary schools.
Districts Should Ensure That Their STEM Curricula Are Focused on the Most Important Topics in Each Discipline, Are Rigorous, and Are Articulated as a Sequence of Topics and Performances	4. **Adoption of instructional materials in grades K-12 that embody the *Common Core State Standards for Mathematics* and *A Framework for K-12 Science Education.****
	5. **Classroom coverage of content and practices in the *Common Core State Standards* and *A Framework for K-12 Science Education.***
Districts Need to Enhance the Capacity of K-12 Teachers	6. **Teachers' science and mathematics content knowledge for teaching.**
	7. Teachers' participation in STEM-specific professional development activities.
Districts Should Provide Instructional Leaders with Professional Development That Helps Them to Create the School Conditions That Appear to Support Student Achievement	8. Instructional leaders' participation in professional development on creating conditions that support STEM learning.
Policy Makers at the National, State, and Local Levels Should Elevate Science to the Same Level of Importance as Reading and Mathematics	9. **Inclusion of science in federal and state accountability systems.**
	10. Inclusion of science in major federal K-12 education initiatives.
	11. State and district staff dedicated to supporting science instruction.
States and National Organizations Should Develop Effective Systems of Assessment That Are Aligned with A Framework for K-12 Science Education *and That Emphasize Science Practices Rather Than Mere Factual Recall*	12. States' use of assessments that measure the core concepts and practices of science and mathematics disciplines.
National and State Policy Makers Should Invest in a Coherent, Focused, and Sustained Set of Supports for STEM Teachers	13. State and federal expenditures dedicated to improving the K-12 STEM teaching workforce.
Federal Agencies Should Support Research That Disentangles the Effects of School Practice from Student Selection, Recognizes the Importance of Contextual Variables, and Allows for Longitudinal Assessments of Student Outcomes	14. **Federal funding for the research identified in *Successful K-12 STEM Education.***

*Because the *Next Generation Science Standards* had not been published at the time of this report, the committee used *A Framework for K-12 Science Education* (National Research Council, 2012) to develop Indicators 4, 5, and 12. These indicators can be tracked in relation to the *Next Generation Science Standards* when they are published.

Data for most of these 14 indicators are, or could be, available through existing surveys administered by the National Center for Education Statistics, although those data sources have limitations that should be considered in light of the goals of the proposed monitoring system. Several of the indicators require new kinds of data collection, changes in the frequency of data collection, or additional research and conceptual development.

A monitoring and reporting system designed around these indicators would be unique in its focus on key aspects of teaching and learning and could enable education leaders, researchers, and policy makers to better understand and improve national, state, and local STEM education for all students. Congress, the National Science Foundation, and the U.S. Department of Education could take the following steps to create such a system:

- Determine whether to create a dedicated survey or use existing federal surveys to collect data on the proposed indicators.
- More fully develop Indicators 1-14, for example, by more precisely defining what the indicators include, identifying what constitutes quality for each indicator, and identifying the most appropriate sources of data.
- Compile, analyze, and report on data that already exist.
- Modify existing surveys or create new data collection mechanisms to yield the needed information.
- Produce regular reports on K-12 STEM education that analyze progress toward the indicators and goals for STEM education.
- Engage stakeholders in discussions of the development of the indicators, their results, and their ongoing utility.

INTRODUCTION

Recent attention to K-12 education in science, technology, engineering, and mathematics (the disciplines collectively referred to as STEM) has revealed challenges in students' performance and persistence, particularly for groups that are underrepresented in the STEM fields (Schmidt, 2011; President's Council of Advisors on Science and Technology, 2010; Lowell et al., 2009; Hill et al., 2008; Higher Education Research Institute, 2010). Although these challenges are daunting, recent education policy developments are creating an unprecedented opportunity to address them.

For example, educational reforms across the country are emphasizing more rigorous common state standards and assessments for all students; increases in school and teacher effectiveness; innovations in teacher preparation and professional development; and new approaches to holding districts, schools, and teachers accountable for results. In addition, the new *Common Core State Standards for Mathematics* (see National Governors Association and Council of Chief State School Officers, 2010) and *A Framework for K-12 Science Education* (National Research Council, 2012)[1] emphasize conceptual understanding of key ideas in each discipline, greater coherence across grade levels, and the practices of science and mathematics. Together, these changes have the potential to engage students in ways that better prepare them for postsecondary study and STEM careers, and thus eventually, for addressing current and future societal challenges and participating in an increasingly global and technologically driven society. The political will and momentum gathering behind these efforts offer an opportunity to realize improvements to K-12 science and mathematics education that have so far remained elusive.

The success of these efforts depends on many factors, including students' equitable access to challenging learning opportunities and instructional materials, teachers' capacity to use those opportunities and materials well, and policies and structures that support effective educational practices. In turn, making informed decisions about improvements to education in STEM requires research and data about the content and quality of the curriculum, teachers' content knowledge, and the use of instructional practices that have been shown to improve outcomes. However, large-scale data are not available in a readily accessible form, mostly because state and federal data systems provide information about *schools* (personnel, organization, and enrollment) rather than *schooling* (key elements of the learning process).

[1]Because the *Next Generation Science Standards* were under development at the time of the report, the committee used the basis for those standards—*A Framework for K-12 Science Education*—to inform this report.

By providing a focused set of key indicators about schooling—students' access to quality learning, educators' capacity, and policy and funding initiatives in STEM—this report addresses the need for research and data that can be used to monitor progress in K-12 STEM education and make informed decisions about improving it. It provides a framework for Congress and relevant federal agencies to create and implement a national-level monitoring and reporting system with the capability to:

- assess progress toward key improvements recommended in the 2011 National Research Council report *Successful K-12 STEM Education*;
- measure student knowledge, interest, and participation in the STEM disciplines and STEM-related activities;
- track financial, human capital, and material investments in K-12 STEM education at the federal, state, and local levels;
- provide information about the capabilities of the STEM-education workforce, including teachers and principals; and
- facilitate strategic planning for federal investments in STEM education and workforce development, when used with labor force projections.

FOUNDATION FOR THIS STUDY

As part of national-level efforts to address the challenges facing K-12 education in STEM, a 2011 report from the National Research Council (NRC), *Successful K-12 STEM Education: Identifying Effective Approaches in Science, Technology, Engineering, and Mathematics,* described three goals for U.S. K-12 education in the STEM disciplines (pp. 4-5):

GOAL 1. Expand the number of students who ultimately pursue advanced degrees and careers in STEM fields and broaden the participation of women and minorities in those fields. This goal focuses on the flow of students to STEM majors and careers as scientists and engineers.

GOAL 2. Expand the STEM-capable workforce and broaden the participation of women and minorities in that workforce. STEM-related careers—including medical assistants and computer and energy technicians—are an increasingly significant part of the U.S. economy (Carnevale, Smith, and Melton, 2011). Most of these careers require an associate degree or vocational certification with specialized STEM knowledge, rather than a bachelor's degree.

GOAL 3. Increase STEM literacy for all students, including those who do not pursue STEM-related careers or additional study in the STEM disciplines. Another goal of education in STEM is to increase students' knowledge and understanding of scientific and mathematical concepts and processes required for personal decision making, participation in civic and cultural affairs, and economic productivity (National Research Council, 1996).

The 2011 report also identified key elements that would be needed to support progress toward these goals: a coherent set of standards and curriculum, teachers with high capacity to teach in their discipline, a supportive assessment and accountability system, adequate instructional time, and students' equal access to high-quality learning opportunities. At the school and district levels, the report recommends specific actions that education leaders and policy makers can take to ensure that these key elements are in place:

- Consider a variety of STEM-focused schools and programs.
- Devote adequate instructional time and resources for science, especially in grades K-5.
- Ensure that curricula in the STEM disciplines are focused on the most important topics in each discipline, are rigorous, and are articulated over time as a sequence of topics and performances.
- Enhance the capacity of K-12 teachers to teach in the STEM disciplines.
- Provide instructional leaders with professional development to create school conditions that support student achievement.

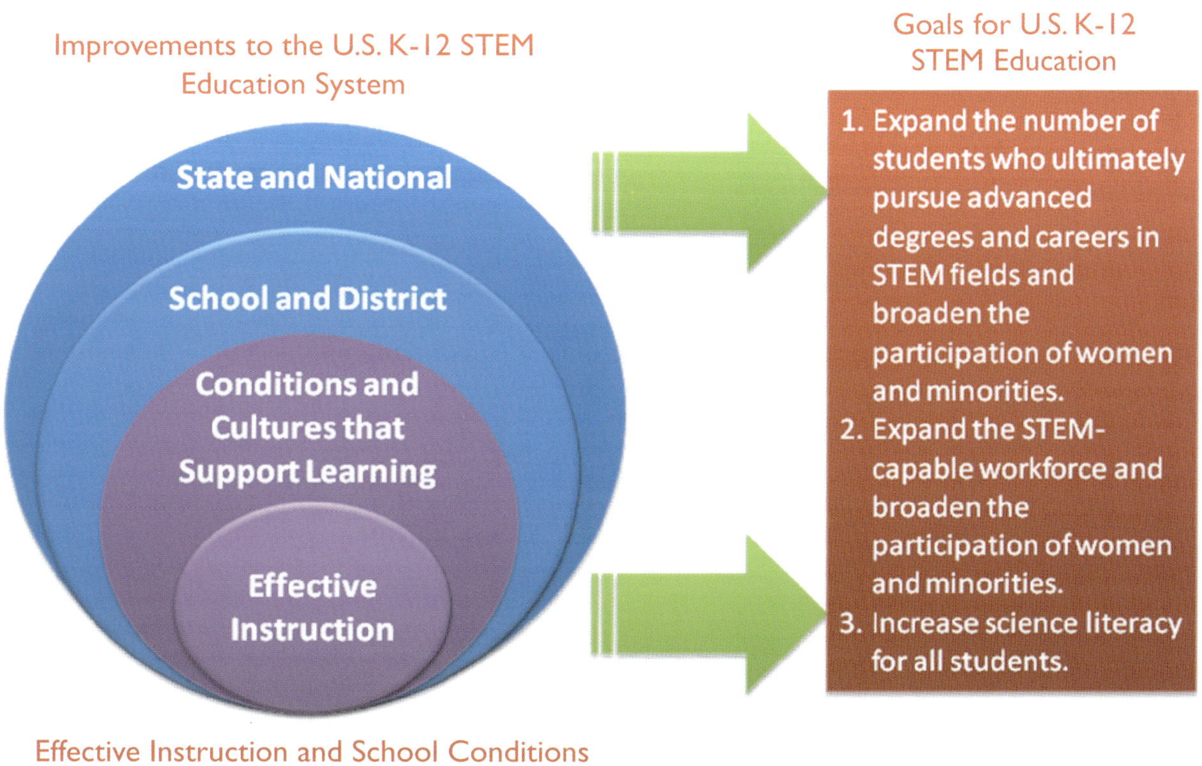

FIGURE 1

Key elements for improvements and goals in *Successful K-12 STEM Education* (National Research Council, 2011).

As shown in Figure 1, to support these changes at the local level, the previous report (National Research Council, 2011) also recommended commensurate enhancements to the national and state infrastructures:

- Elevate science to the same level of importance as reading and mathematics.
- Develop effective systems of assessment for science.
- Invest in supports for teachers in the STEM disciplines.
- Support rigorous research to identify instructional practices that improve student outcomes.

STUDY OVERVIEW

Because education in the United States is decentralized, it is likely that schools, districts, and states will implement the recommendations in *Successful K-12 STEM Education* in varied ways and to varying degrees. And although some data that could provide indications of progress in meeting the recommendations of the 2011 report are currently collected at the state and national levels, these data are not collected or analyzed expressly for that purpose, nor is there any attempt to coordinate data collection. As a result, educators, researchers, and policy makers lack a clear picture of the extent to which the nation as a whole is implementing the recommended improvements to STEM education, or making progress toward them.

To address this lack, and in response to a request from Congress, the National Science Foundation (NSF) requested that the National Research Council's Board on Science Education and Board on Testing and Assessment convene an ad hoc committee (see Box 1 for the formal charge to the committee). This report of the Committee on the Evaluation Framework for Successful K-12 STEM Education presents the committee's proposed indicators and framework for creating a monitoring and reporting system with those indicators at its core.

Scope of the Study

The committee's charge addresses only the recommendations in *Successful K-12 STEM Education*. However, those recommendations were developed to support improvements that would enable progress toward the broader goals of U.S. STEM education (see Figure 1). Thus, the committee decided that it is important for this proposed monitoring and reporting system to include information related to the goals for STEM education. By including indicators of conditions that lead to improved performance of students and the education system alongside key performance measures that can be tracked over time to evaluate whether the United States is making progress toward its STEM education goals, the committee's proposed monitoring and reporting system has the potential to support better decision making and contribute to the desired improvements (Walpole and Noeth, 2002).

Reflecting the emphasis in *Successful K-12 STEM Education*, this report primarily addresses science and mathematics education rather than all of the STEM disciplines. Because the research base that underpins *Successful K-12 STEM Education* was the strongest for mathematics and science, the recommendations in that report relate mostly to those subjects, with some of the recommendations addressing only science. Research in technology and engineering education is less mature because those subjects are not as commonly taught in K-12 education (National Research Council, 2011, p. 2). However, engineering and technology are included in *A Framework for K-12 Science Education* (National Research Council, 2012), which is forming the basis for the *Next Generation Science Standards* (being developed by Achieve, Inc.). Thus, any of the committee's proposed indicators that relate to the 2012 report include engineering and technology. In addition, the scheduled 2014 National Assessment of Educational Progress (NAEP) technology

BOX 1
Charge to the Committee

An ad hoc committee will conduct a fast-track, targeted consensus study to identify methods for tracking and evaluating the implementation of the improvements for K-12 education recommended in the 2011 NRC report, *Successful K-12 STEM Education*. At the school and district levels, these recommended improvements include adequate instructional time and resources for science, coherent standards and curriculum, greater teacher capacity, and supportive school conditions that have been identified in the research. At the state and national levels, these recommended improvements include greater attention to science, including assessments for science, investment in support for STEM teachers, and increased support for research programs that can identify instructional practices that improve student outcomes in STEM.

The committee was charged with authoring a short report that includes a plan for evaluating progress toward the 2011 report's recommendations. The short report is to provide guidance to enable NSF and Congress to consider how to support a full-scale evaluation in the future. To that end, the report will

- identify existing and additional measures needed for tracking the improvements recommended in 2011,
- discuss the feasibility of including such measures in existing programs of data collection and linking them to existing and improved measures of science learning outcomes,
- identify additional research that would be needed to inform a full-scale evaluation of progress toward the 2011 report's recommendations, and
- make new recommendations regarding the roles of various federal and state institutions in supporting the needed research and data collection for an evaluation of progress.

and engineering framework is increasing the emphasis on these subjects by measuring students' literacy in them. As these subjects become more central to K-12 education in the future, they can be more fully incorporated into the proposed monitoring system, with additional research to identify the most important aspects to measure.

Study Approach and Sources of Evidence

The committee worked through an iterative process of gathering information, deliberating, identifying gaps and questions, gathering further information to fill those gaps, and further deliberations. The recommendations in *Successful K-12 STEM Education* (National Research Council, 2011) were the starting point for our deliberations: we took the previous committee's work and its recommendations as given and did not debate their merit. To identify indicators of progress that would be important to measure for each recommendation, we also drew on a long series of research reports related to K-12 science and mathematics instructional practices (National Research Council, 1999, 2001, 2006, 2007, 2009, 2012; National Academy of Engineering, 2010); school conditions, leadership, and teacher capacity building (e.g., Bryk et al., 2010; Gamoran et al., 2003; McLaughlin and Talbert, 2006); learning science in informal environments (National Research Council, 2009); and general principles of monitoring and evaluation (Shadish, Cook, and Campbell, 2001; U.S. General Accounting Office, 2003). Finally, we consulted relevant experts at agencies that manage large-scale data sets (e.g., National Center for Education Statistics, National Science Foundation) and other organizations (e.g., Council of Chief State School Officers) about the availability of certain types of data.

The rest of this report identifies key indicators that could provide evidence of progress toward the recommendations in *Successful K-12 STEM Education*. We set four criteria to guide the selection of indicators. Indicators must

- be shown by research to be related to a given recommendation,
- generate information that clearly enhances understanding of progress toward the recommendation for all student groups,
- have the potential to be used for continuous improvement, and
- be feasible to measure on a large scale.

Some of the proposed indicators require new kinds of data collection, and all would benefit from additional research and conceptual development to enhance understanding of what to measure. We included indicators as long as they met the above criteria, but we did not undertake a systematic analysis of the costs involved in measuring the proposed indicators. Instead, we drew on the committee's expertise with federal research and data collection systems and on consultations with relevant experts to determine whether collecting data on the proposed indicators would generally be feasible.

The monitoring and reporting system proposed in this report is not the first of its kind. In the late 1980s similar indicator systems were developed for K-12 science and mathematics education (National Research Council, 1988; Shavelson, McDonnell, and Oakes, 1989), and the committee consulted those reports. The indicators in those reports were comprehensive, were carefully developed, and had the potential to generate valuable information about the condition of mathematics and science education in the nation's schools and how individual components of the education system interact to affect the system as a whole (National Research Council, 1988; Shavelson, McDonnell, and Oakes, 1989). Although they were never fully implemented, these indicator systems provided a base for our work.

As we note at the beginning of this report, the committee's proposed indicators are being offered at a time of promising changes to what is taught in K-12 science and mathematics, how students are assessed in those subjects, and how teachers are prepared to meet those changing demands. Although the proposed indicators overlap to some degree with previously developed indicators, they reflect current national priorities and recent research on teaching and learning in STEM. They also reflect the continuing interest of Congress and NSF in putting the recommendations of *Successful K-12 STEM Education* into action. By proposing a comparatively small set of focused, specific, and actionable indicators that are driven by research on impact, the committee hopes that the current political will and momentum will be harnessed to build and sustain a valuable monitoring and reporting system.

INDICATORS FOR MEASURING IMPROVEMENTS TO THE U.S. K-12 STEM EDUCATION SYSTEM

In this section, we identify and describe a set of indicators that could provide evidence of progress toward each recommendation of *Successful K-12 STEM Education*. Although the indicators are linked to specific recommendations in that report, taken together, they address several key elements of successful K-12 education in STEM: access to quality learning opportunities, educators with high capacity to teach in their disciplines, and supportive policies and funding initiatives. Thus, the committee's intent is for these indicators to form the core of a national program to monitor the health of the education system in STEM. It is not the committee's aim for these indicators to become part of a new accountability system for K-12 education in STEM. Rather, the goal is for the National Center for Education Statistics (NCES) and NSF to generate information that enables education leaders, researchers, and policy makers to understand and improve state and local education systems.

We propose indicators on five topics related to recommendations for school districts in the 2011 report:

- multiple models of STEM-focused schools,
- adequate instructional time and resources for science in grades K-5,
- high-quality curricula,
- the capacity of K-12 teachers, and
- professional development for instructional leaders.

We also propose indicators on four topics related to recommendations for state and national policy makers in the 2011 report:

- elevated status for science;
- effective systems of assessment;
- federal and state support for STEM teachers; and
- research to enhance understanding of STEM schools, practices, and outcomes.

For each indicator, we discuss available and potentially available data, with "potentially available" defined as data that could be collected by modifying existing data collection systems (see the Appendix for information about surveys that are the potential sources of data). We also discuss data and research needs for each proposed indicator.

The proposed indicator system relies heavily on the NCES Schools and Staffing Survey (SASS).[2] Making all of the proposed modifications to SASS might have negative effects on response rates if

[2]When this report was written, the most recently available SASS was 2007-2008.

the survey becomes too onerous for respondents. Thus, any modifications would ideally be undertaken as part of efforts to systematically streamline SASS. More broadly, placing the burden of this monitoring system on the shoulders of existing national surveys will require strategic decisions about the frequency, subject matter, and question rotations of those surveys.

Developing new kinds of data collection for some of the indicators might help to alleviate this problem. For example, background questionnaires for students, teachers, and schools that could be administered with assessments related to the *Common Core State Standards for Mathematics* (and eventually, *A Framework for K-12 Science Education*[3]) would be a valuable new data collection mechanism for several of the proposed indicators. Because student assessments aligned with the *Common Core State Standards for Mathematics* are currently under development, an exceptional opportunity exists to develop accompanying background surveys that directly measure the key elements of these reforms. Such surveys could become the primary data collection vehicle for several of the proposed indicators. They would be especially valuable because they would be regularly administered across the majority of states and could be coupled with student achievement data.

Although the proposed indicators do not specifically mention different student populations, the aim of equitable access to resources and learning opportunities for all students is central to the goals for education in STEM. Because disparities in access to high-quality learning opportunities, instructional materials, and teachers contribute to achievement gaps among students from different racial, ethnic, language, and socioeconomic groups, tracking patterns in access to those resources is an essential component of Indicators 1-8. Thus, the committee's intent is for data on Indicators 1-8 to be collected and analyzed in a manner that provides an understanding of variation among different student populations and socioeconomic contexts.

Multiple Models of STEM-Focused Schools

As noted in *Successful K-12 STEM Education,* high-quality education in the STEM disciplines can take place in diverse public school settings, including STEM-focused schools with selective admission policies, STEM-focused schools with inclusive admission policies, STEM-focused career and technical education programs, and comprehensive public schools (see National Research Council, 2011, for a more complete description of the school types). Because these schools often have different goals and pursue different strategies to meet those goals, and because the evidence is not sufficient to recommend one type of school over another, *Successful K-12 STEM Education* recommended that "districts seeking to improve STEM outcomes beyond comprehensive schools should consider all three models of STEM-focused schools" (National Research Council, 2011, p. 27). Although variation exists within and across these categories, they share an emphasis on the STEM

[3]Because the *Next Generation Science Standards* were under development at the time of the report, the committee used the basis for those standards—*A Framework for K-12 Science Education*—to inform this report.

disciplines, and on providing access to courses and experiences that will prepare students to be scientifically literate and perhaps pursue careers or further study in the STEM disciplines after high school (National Research Council, 2011).

Key Indicator to Monitor

As a first step toward measuring progress toward this recommendation, the committee proposes collecting descriptive information to quantify the availability of STEM-focused schools and programs. The indicator that we propose below is a measure only of quantity—and thus, the degree of access to STEM-related learning experiences. To support decisions about the types of schools or programs in which districts should invest, eventually it will be necessary to collect data that address the quality of these schools and programs.

INDICATOR 1. Number of, and enrollment in, different types of STEM schools and programs in each district.

This indicator is intended to measure the extent to which all students have the opportunity to pursue some kind of focused experience in STEM as part of their K-12 education, which is particularly important for students in areas with limited resources. The indicator should include selective STEM schools, inclusive STEM schools, STEM-focused career and technical education schools or programs, and STEM-focused programs in comprehensive schools, as defined in *Successful K-12 STEM Education*. Public charter and noncharter schools of these types should be included in the monitoring system.

To qualify as a specialized STEM experience, a school or program would need to provide all of its students with a range and depth of STEM learning experiences that exceed state requirements. Developing the criteria for each type of STEM-focused school or program might involve an expert meeting with individuals who have experience and expertise in implementing and studying such programs.

Available and potentially available data. Currently, the primary way to count STEM-focused schools is by searching databases of school names for the words science, technology, engineering, or mathematics. This method does not provide a full or accurate count of STEM-focused schools, in part because there is no uniform definition of what constitutes a "STEM-focused" school or program.

Surveys conducted by NCES (e.g., the SASS, the High School Longitudinal Study, and the National Education Longitudinal Study) include yes/no questions about whether a school has a special program emphasis, but do not differentiate among different themes. The SASS also contains questions about career and technical education schools and programs, but they, too, are not differentiated to focus on the STEM disciplines. Additional questions could be added to the SASS or to the National Civil Rights Data Collection as a way of identifying various types of STEM-focused schools and programs. However, these data sets do not contain a census of all schools or use a common definition of "STEM-focused." To provide a census, this information could be added to the NCES Common Core of Data. Regardless of the survey used, measuring this indicator would involve creating definitions of the elements that characterize a STEM-focused school and would require districts and states to consent to using those common definitions in their reporting.

Data and research needs. Research is needed to define the criteria for STEM-focused schools and programs and use them as the framework for developing survey items for administration to school principals. Surveys are needed that elicit principals' reports of the requirements for entering their STEM-focused school or program, to ascertain the extent to which the program targets those who have already demonstrated STEM talent. These surveys also should capture the geographic areas from which the school draws its students so that analysts can derive estimates of the extent to which different student subgroups do or do not have equal access to the opportunities. Further research also is needed on the essential characteristics of effective STEM schools or programs, leading to the development of indicators that assess whether schools or programs within schools have these characteristics.

Adequate Instructional Time and Resources for Science in Grades K-5

The recommendation in *Successful K-12 STEM Education* that "districts should devote adequate time and resources to science education in grades K-5" (National Research Council, 2011, p. 27) was proposed to mitigate an unintended consequence of the federal accountability system in education: that instructional time for science in elementary school has been reduced to devote additional time to reading and mathematics (Center on Education Policy, 2007; Dorph et al., 2011). Reducing the time devoted to science in the elementary grades is of special concern because some research suggests that "life experiences before 8th grade and in elementary school may have an important impact on future career plans," which requires "close attention to children's early exposure to science" (Tai et al., 2006, p. 1144).

Key Indicators to Monitor

To the extent that early experiences in science are valuable in preparing students for future science learning and careers, reducing the exposure to science in elementary school is particularly problematic for students who do not have access to science learning opportunities in their homes and communities (National Research Council, 2007). For these students, limiting early science learning opportunities leaves them unprepared for science courses in middle and high school (Hartry et al., 2012), which can exacerbate future inequities in interest, course-taking, and achievement, in STEM.

The committee proposes two indicators related to instructional time. First, as a proxy for the value that is placed on science, it is essential to measure the number of instructional minutes allocated to science. It also is important to consider the characteristics of that time. For example, implementing instructional approaches that afford students opportunities to engage in the practices of science requires more time than the 30-45 minute session that elementary schools typically allocate to a science lesson. Teachers who have the flexibility to consider time on a weekly basis may create a larger block of time to allow students to plan and carry out investigations, construct explanations, and engage in building an argument from evidence (Dorph et al., 2011).

Second, when measuring time devoted to science, it is also important to include opportunities that schools provide for students to engage in science learning both in and beyond formal class time (OECD, 2012). These opportunities vary widely and can include field trips to local science-rich institutions, as well as after-school programs, science camps, clubs, and competitions. Opportunities for science learning in and outside the classroom are particularly important for students who do not have access to science learning opportunities beyond the school (National Research Council, 2009).[4]

The committee did not attempt to define what was meant by "adequate" time and resources, or to measure the quality of instructional time. Rather, Indicators 2 and 3 are intended to provide ongoing measures of the amount of time and the kinds of opportunities that are available for science learning in the elementary grades. Determining adequacy would be considerably more difficult and might entail analyses of the quality of instructional time and the relationship between the average amount of time devoted to science instruction and the level of demand of a state's science standards. It also might entail additional research on any learning loss from time not spent on other subjects and on the relative effectiveness of different organizational structures for instructional time.

INDICATOR 2. Time allocated to teach science in grades K-5.

This indicator simply measures teachers' estimates of the amount of time that they devote to teaching science. Time should include the number of instructional minutes per week, as well as the different configurations elementary teachers use to implement those instructional minutes.

Available and potentially available data. The SASS teacher questionnaires include questions about time allocated to general subjects such as science and mathematics, as does the NAEP grade 4 teacher survey. Although NAEP offers the opportunity to link survey results with student achievement data, the response options may require modification to accurately assess time use. These limitations notwithstanding, estimates from the existing SASS and NAEP surveys would

[4]Although the original recommendation addressed instructional time and resources for science in grades K-5, Indicators 2 and 3 only address time. Indicator 4 addresses instructional resources.

allow for state-level indicators or for comparisons of schools that serve different demographic populations. The surveys could be amended to include questions about the extent to which instructional time that is devoted to other subjects includes science content.

Data and research needs. Although data are available for this indicator, existing measures rely solely on teacher self-reported data. Additional research is needed to assess the reliability of these indicators, for example by comparing teachers' self-reports of time usage with data from classroom observations and teacher logs. Additional research also might be needed to determine how to measure time devoted to science when science is taught in the context of other subjects, perhaps building on the work of Dorph et al. (2011) and Hartry et al. (2012).

INDICATOR 3. Science-related learning opportunities in elementary schools.

This indicator is intended to reflect the full range of science learning opportunities that elementary schools provide for students. It should include a focused examination of in-school but non-classroom science learning experiences, together with out-of-school opportunities that schools and districts intentionally provide to enhance their science offerings for all students. The latter may include science centers, museums, zoos, or STEM-related businesses and may depend on the access that different communities have to such science-rich institutions and resources.

Available and potentially available data. The NCES High School Longitudinal Study includes questions about high school students' participation in science, engineering, and mathematics competitions, museums, clubs, and extracurricular activities. Similar items are under consideration for the kindergarten cohort study. If those questions are asked in the earlier grades, they could be modified to indicate whether these opportunities are offered through the school. The OECD's Programme for International Student Assessment (PISA) is developing teacher questionnaires for the 2015 science-focused assessment that also could be modified or adapted. A set of relevant questions also could be added to the SASS teacher survey or the NAEP science teacher survey, although the latter would provide information only for grade 4.

Focused, Rigorous, and Sequenced Curricula

Typical science and mathematics curricula in the United States have been criticized as being fragmented and containing too much material for students to be able to build understanding over time (Valverde et al., 2002; Schmidt et al., 2001; Schmidt, McKnight, and Raizen, 1996). In response to this concern, *Successful K-12 STEM Education* recommended that "districts should ensure that their STEM curricula are focused on the most important topics in each discipline, are rigorous, and are articulated as a sequence of topics and performances" (National Research Council, 2011, p. 27).

The *Common Core State Standards for Mathematics* (National Governors Association and Council of Chief State School Officers, 2010) and *A Framework for K-12 Science Education* (National Research Council, 2012) were designed to address these concerns. The *Common Core State Standards for Mathematics* have been adopted by 45 states, and *A Framework for K-12 Science Education* is forming the basis for the *Next Generation Science Standards* that are currently under development by Achieve, Inc. As significant drivers of K-12 mathematics and science education, these documents provided the context for this committee's work; however, the following indicators are framed in such a way that relevant data also can be collected in states that do not adopt the standards.

The quality of the standards and the effects of adopting them have not yet been fully evaluated. As more research becomes available about these important issues, the proposed indicators that are linked to the standards will need to be revisited and refined.

Key Indicators to Monitor

In the strictest sense, measuring progress toward this recommendation of *Successful K-12 STEM Education* would entail determining how many districts had adopted curricula that embody the *Common Core State Standards for Mathematics* and *A Framework for K-12 Science Education* or were solidly grounded in current research on teaching and learning in science and mathematics. Although it would be useful to know which curricula are being used and the extent to which they embody research on learning, simply adopting focused, rigorous, and coherent curricula is not sufficient to improve instruction and student outcomes. Thus, the committee also proposes an indicator related to teachers' reports of how those curricula are being implemented.

Indicators that are related to curriculum may include engineering, as well as career and technical education. Career and technical education is a potentially important pathway to prepare students for STEM-related careers, including those in the information technology, computer science, and health fields (Silverberg et al., 2004; National Research Center for Career and Technical Education, 2010). A limited amount of evidence suggests that career and technical education, "assumed to motivate learning through real-life applications, does not have to be in conflict with academic achievement" (National Research Council, 2011, p. 13), as long as the curricula integrate rigorous academic content with occupational training (Stone, Alfeld, and Pearson, 2008).

As this report was being written, curricular materials that embody the *Common Core State Standards for Mathematics* were not in widespread use, nor were those that embody *A Framework for K-12 Science Education*.[5] Thus, Indicators 4 and 5 cannot be monitored until it is clear what materials have been

---

[5]Because the *Next Generation Science Standards* had not been published at the time of this report, the committee used *A Framework for K-12 Science Education* (National Research Council, 2012) to develop Indicators 4, 5, and 12. These indicators can be tracked in relation to the *Next Generation Science Standards* when they are published.

developed, which districts have adopted particular curricular materials and assessments, and how they are using those resources. Nonetheless, it is advisable to develop the indicators before those materials are available so their use can be monitored from the outset.

INDICATOR 4. Adoption of instructional materials in grades K-12 that embody the *Common Core State Standards for Mathematics* and *A Framework for K-12 Science Education.*

This indicator would provide descriptive information about which districts have adopted instructional materials that embody the *Common Core State Standards for Mathematics* and *A Framework for K-12 Science Education* or that have been shown by research to improve student achievement and proficiency with the practices of science or mathematics.

The committee proposes a two-tiered data collection for this indicator. The first tier includes determining which curricula districts and schools have adopted for science, mathematics, engineering, and career and technical education. The second tier involves analysis by an independent entity of the extent to which the most widely used curricula include the practices of science and mathematics, as specified in the *Common Core State Standards for Mathematics* and *A Framework for K-12 Science Education*.

Available and potentially available data. No data are currently collected that would provide information on a large scale about this indicator. For general K-12 mathematics, science, and engineering education, some questions might be added to the SASS for teachers to ascertain which instructional materials—main and supplemental—they use. Questions about district-level adoption of curricula could be added to a district-level survey such as the National Civil Rights Data Collection. These questions also could be incorporated into questionnaires that might accompany the assessments that are eventually developed in conjunction with mathematics and science standards. Regarding the alignment of career and technical education materials to standards, the National Center for Research on Career and Technical Education, and some individual states, are undertaking efforts to map career and technical education curricula to the *Common Core State Standards for Mathematics*.

Data and research needs. Considerable research and development work is needed to create a transparent, scientifically neutral set of criteria for determining the degree to which the most widely used instructional materials embody the standards or have been shown by research to improve student achievement and proficiency with the practices of science, mathematics, or engineering. Ideally, the reviews of curricular materials would be available as the materials were produced so that districts could use the reviews to inform their purchasing decisions.

Research to develop these criteria could build on current efforts to map career and technical education curricula to standards, as well as previous efforts by the American Association for the Advancement of Science (2005) in science and Gueudet, Pepin, and Trouche (2012) and Schmidt et al. (1997) in mathematics. Most of these efforts have addressed content; less research has been done to analyze the extent to which curricula address the practices of science, mathematics, and engineering (e.g., building an argument from evidence, constructing explanations, and designing solutions). Much work remains to define these practices and determine how they might look across different grades and subject areas. Such work may include pilot studies to develop coding schemes for assessing curricula.

INDICATOR 5. Classroom coverage of content and practices in the *Common Core State Standards for Mathematics* and *A Framework for K-12 Science Education*.

The opportunity students have to learn content and practices is a critical indicator that has been shown in numerous studies to be related to achievement and distributed inequitably across different populations of students (Schmidt and Maier, 2009; Schmidt and McKnight, 2012). Content coverage (or opportunity to learn) is defined in three ways: (1) the extent of coverage, (2) the amount of time devoted to content, and (3) the order of coverage. Because it is central to academic achievement, coverage provides an intermediate indicator related to the quality of schooling. In fact, in many countries, content coverage is one of the important judgments usually made by inspectors. Here we propose it as a statistical indicator based on teacher responses to survey items.

Available and potentially available data. Regarding nationally collected data, the SASS includes questions about time allocated to general subjects such as science and mathematics. The survey could be amended to add questions about specific topics taught in each subject, linked to the core ideas and the mathematical, scientific, and engineering practices presented in the *Common Core State Standards for Mathematics* (National Governors Association and Council of Chief State School Officers, 2010) and *A Framework for K-12 Science Education* (National Research Council, 2012). However, doing so likely would increase respondent burden, and it might require subsampling mathematics and science teachers in particular grades.

The existing NAEP science teacher survey includes some items that map well to the practices of science that are described in *A Framework for K-12 Science Education*. Data from these questions could be used immediately to provide baseline measurement. These surveys could also be modified and expanded to provide more thorough and comprehensive measures of classroom coverage of the content and practices of science and engineering. Appropriate questions about classroom coverage of science and mathematics content and practices also could be added to future surveys that are developed in conjunction with new assessments in science and mathematics.

Several other large-scale efforts have examined classroom coverage of science and mathematics content and have used other measures to validate teachers' self-reported data. For example, the Surveys of Enacted Curriculum, which are used by hundreds of schools across the country, examine the content and cognitive demand of K-12 mathematics, science, and English language arts curriculum and could serve as a baseline for examining instructional change over time (Porter, Polikoff, and Smithson, 2009; Porter et al., 2007; Smithson and Blank, 2006). Although those surveys were not designed to be specifically tied to current standards documents, they could be adapted for that purpose. A different, large-scale effort to measure classroom coverage in Ohio and Michigan (Schmidt and McKnight, 2012) has used a Web-based survey of classroom coverage that was based on the international instruments developed in the Trends in International Mathematics and Science Study (TIMSS) and modified to reflect the *Common Core State Standards for Mathematics*.

Data and research needs. Current efforts to measure classroom coverage are more developed for mathematics than for science, engineering, or career and technical education. To date, these efforts have focused more on measuring content than on the standards for mathematical practice (e.g., modeling with mathematics, reasoning abstractly and quantitatively) or the practices of science and engineering (e.g., analyzing and interpreting data, engaging in argument from evidence). Considerable research and development efforts are needed to develop measures for the coverage of mathematical practices, and for coverage of content and practices in engineering, science, and career and technical education. Such efforts might draw on previous and ongoing surveys (e.g., Porter, Polikoff, and Smithson, 2009; Schmidt and McKnight, 2012; Dorph et al., 2011; and Smith et al., 2002), while addressing well-documented concerns about the validity of teacher self-report data (Hudson, McMahon, and Overstreet, 2002).

Enhanced Capacity of K-12 Teachers

Teaching in ways that inspire students and deepen their understanding of science and mathematics content and practices requires teachers to have content knowledge and expertise in teaching that content (National Research Council, 2010). *Successful K-12 STEM Education* recommended that "districts need to enhance the capacity of K-12 teachers" (National Research Council, 2011, p. 27), in part because many current teachers, including those who are teaching out of their field of expertise, are underprepared for the demands of STEM teaching. However, professional development in science and mathematics, when available, is often short, fragmented, not designed to address the specific needs of individual teachers, and therefore ineffective (National Research Council, 2011, pp. 20-21). Developing the requisite knowledge and teaching strategies will require profound changes to current systems for supporting teachers' learning across their careers.

Key Indicators to Monitor

Indicators to measure progress toward this recommendation of *Successful K-12 STEM Education* include baseline information about teachers' content knowledge and knowledge of how to effectively teach science or mathematics (pedagogical content knowledge) and information about participation in high-quality professional development activities.

INDICATOR 6. Teachers' science and mathematics content knowledge for teaching.

Content knowledge and knowledge of effective teaching and learning strategies are important components of teaching effectively. Teachers must integrate content and pedagogy in ways that reflect an understanding of "how students' learning develops in [a given] field, the kinds of misconceptions students may develop, and strategies for addressing students' evolving needs" (National Research Council, 2010, p. 73). Thus, this indicator is intended to measure the depth of teachers' understanding of the content they teach, as well as the knowledge needed for effective teaching (e.g., common misunderstandings that students have about a topic).

Available and potentially available data. Teachers' degrees and the courses they take in college often are used as proxies for science or mathematics content knowledge. Other measures are needed, however, because research has shown that these indicators do not consistently predict student achievement at the secondary level (Wilson, Floden, and Ferrini-Mundy, 2001). In contrast, teachers' own reports of their capacity to teach certain topics have been reliably shown to indicate content knowledge, as long as "stakes" are not attached to teachers' responses (PROM/SE, 2006).

Questions about teachers' perceived capacity to teach certain topics in mathematics appear on the TIMSS surveys for grades 4 and 8. These questions could be amended so that they embody the key tenets of the *Common Core State Standards for Mathematics* and *A Framework for K-12 Science Education* (National Research Council, 2012). Questions that are linked to the core ideas in those documents also could be added to teacher questionnaires in the longitudinal studies conducted by NCES. Alternatively, such questions could be added to the SASS, or for grade 8 teachers on NAEP. Questions about teachers' self-reported abilities to teach particular content areas also could be added to future surveys that are developed in conjunction with new assessments in mathematics and science. Whatever mechanism is used to collect these data, it will be important to ensure the confidentiality of responses so that teachers do not fear negative consequences for candid self-evaluations.

Regarding more direct measures of teachers' knowledge of mathematics and science for teaching, the Teacher Education and Development Study has measured the level of mathematics and related teaching knowledge that teachers acquire in their preparation programs in 18 countries (Tatto et al., 2008). In addition, the Learning Mathematics for Teaching Project has developed surveys that measure teachers' knowledge for teaching mathematics and it has used those surveys to understand how teachers acquire mathematical knowledge for teaching, and how that knowledge relates to students' achievement in mathematics (Hill, Schilling, and Ball, 2004; Hill and Ball, 2004; Hill, Rowan, and Ball, 2005). Work to develop similar instruments for science is under way (e.g., Smith and Taylor, 2010), but faces challenges such as the need to measure content knowledge for multiple subject areas.

Data and research needs. Although there is widespread agreement that teacher content knowledge for teaching is essential, research on this issue is at an early stage. Expanding the emerging research on direct measures of teacher knowledge in STEM fields is needed, as are measures of teachers' abilities to integrate content knowledge with understanding of student thinking. Because these efforts are more mature for mathematics, greater investments will be needed to develop similar measures for science and engineering.

INDICATOR 7. Teachers' participation in STEM-specific professional development activities.

This indicator is designed to measure teachers' participation in high-quality, research-based professional development in science and mathematics. Although the research is not conclusive, there is emerging consensus that high-quality professional development (a) focuses on developing teachers' capabilities and knowledge to teach content and subject matter, (b) addresses teachers' classroom work and the problems they encounter in their school settings, and (c) provides multiple and sustained opportunities for teacher learning over a substantial time interval (National Research Council, 2011, p. 21). Because the challenges of teaching science and mathematics and of providing quality professional development differ by grade level and subject matter, progress toward this indicator should be tracked along those dimensions.

A desirable goal is for the content of a professional development program in STEM to be rooted in practices that have been found to be effective in studies that use valid scientific designs such as experiments or quasi-experiments. However, a review of 1,343 studies of professional development revealed that only 9 of them had the types of designs—randomized control trials or quasi-experimental designs—that allow causal inferences about the effectiveness of the professional development strategies they examined (Yoon et al., 2007). To assess changes in teachers' instructional practice as well as changes in students' learning, these kinds of studies should be contrasted with "customer satisfaction" reports that often are compiled about whether the audience "liked" the professional development activity or the speaker. Although such satisfaction reports might be appropriate for monitoring whether users believed a given professional development activity was useful for them, they do not convey information about whether the activity was designed around evidence-based practices.

Available and potentially available data. Participation in professional development will be measured in the 2013 OECD Teaching and Learning International Survey, in which the United States will participate. In addition, the SASS asks whether teachers have participated in any professional development specific to and concentrating on the content of the subject(s) they teach. Mathematics and science teachers can be identified in this manner, but the surveys do not provide information about mathematics- or science-specific professional development for teachers who do not designate a single subject as their main teaching assignment (e.g., elementary school teachers). Although it would be possible to restrict the sample from the 2007-2008 SASS to secondary mathematics and science teachers and describe their responses to questions about whether they participated in professional development activities specific to the subjects they teach, the resulting data set would leave the field uninformed about the science professional development received by elementary teachers.

The 2007-2008 SASS asked questions about the duration and perceived value of professional development, and the 2003-2004 SASS asked questions related to professional development that involves peers and shared planning time. Data from the 2003-2004 survey could be compiled to provide some insights into the nature of professional development for mathematics and science teachers, and the relevant questions from that survey could be reused in future versions of the SASS. Survey questions from the evaluation of the Eisenhower Professional Development Program (Garet et al., 2001), ongoing surveys of professional development for mathematics and science teachers (Banilower et al., 2006), and California surveys of science education (Dorph et al., 2011; Hartry et al., 2012) also could be adapted for this purpose. Relevant questions could be added to new questionnaires that might be developed in conjunction with mathematics and science assessments, which would offer the opportunity to link student achievement data to teachers' participation in professional development.

Data and research needs. Although knowledge is accumulating on professional development (Desimone, 2009; Hochberg and Desimone, 2010), additional research is needed to determine the characteristics of effective professional development for science and mathematics teachers (e.g., Garet et al., 2011), and potential obstacles to engaging in professional development. Such research could evaluate the quality of existing professional development activities, the range of opportunities that might lead to improvements in teachers' practice, and the extent to which they are aligned to instructional policies such as the Common Core State Standards. Because research that links professional development activities to student outcomes is especially sparse, additional research in this area would be particularly valuable. A sharper focus on understanding the prevalence of evidence-based practices in professional development also would bolster this indicator; however, no system currently exists for doing so.

Professional Development for Instructional Leaders

Although it is necessary to have qualified, capable teachers, research has shown that school context matters just as much as teachers' qualifications (DeAngelis and Presley, 2011; McLaughlin and Talbert, 2006). Longitudinal research in Chicago elementary schools identified five common elements shared by elementary schools that improve reading and mathematics scores (Bryk et al., 2010): (1) school leadership as the driver for change, (2) professional capacity of school staff, (3) strong ties with parents and the community, (4) a student-centered learning climate, and (5) instructional guidance for teachers. Schools that are strong in these areas are much more likely to improve student learning than schools that are not strong in these areas, and these supports have been associated with improved learning even in neighborhoods of extreme hardship and poverty. The available evidence does not indicate whether these supports also improve science achievement, and whether the same supports are important in middle and high schools, but it is clear that strong school leadership is vital to create school conditions and cultures that support successful education in all subjects (Bryk and Driscoll, 1988; Stein and Nelson, 2003). Thus, *Successful K-12 STEM Education* recommended that "districts should provide instructional leaders with professional development that helps them to create the school conditions that appear to support student achievement" (National Research Council, 2001, p. 27).

Key Indicator to Monitor

The recommendation from *Successful K-12 STEM Education* was intended to ensure that school leaders receive supports to create the school conditions identified above. The committee proposes one indicator to provide descriptive information about the extent to which school principals participate in high-quality, research-based professional development to help them create those conditions. It does not measure the prevalence of the school conditions that support learning, which would require another set of indicators and additional research to develop those indicators.

INDICATOR 8. Instructional leaders' participation in professional development on creating conditions that support STEM learning.

Professional development for leaders should be defined broadly, and could include activities such as coaching and time to discuss work with peers.

Available and potentially available data. The SASS asks about participation in general professional development for school principals, but it does not ask about the characteristics or quality of that professional development or about professional development related to instructional leadership in specific subjects. Data are needed to provide information on program focus, duration, and the presence of features that research has shown to be effective. The SASS could be modified to address specific subjects and the issue of quality of professional development for school leaders. Relevant questions on the 2003-2004 and 2007-2008 SASS for teachers about the nature, duration, and perceived value of professional development also could be adapted for principals. A set of similar questions about school leaders' exposure to professional development also could be added to the science section of the NAEP school questionnaire for administrators, or to questionnaires that might be developed in conjunction with assessments related to the *Common Core State Standards for Mathematics* and *A Framework for K-12 Science Education*.

Data and research needs. Additional research would be needed to identify what effective instructional leadership in mathematics and science might look like (e.g., observation and feedback on instructional practices, observing teachers during collaborative activities, supporting the work of math and science coaches) and then to evaluate the quality of various professional development opportunities that promote effective leadership, similar to work by Cobb et al. (in press). Data on the alignment of the focus of the professional development with STEM disciplines also would be needed.

Elevated Status for Science

The recommendation from *Successful K-12 STEM Education* to "elevate science to the same level of importance as reading and mathematics" (National Research Council, 2011, p. 28) was developed in response to an unintended consequence of the emphasis on mathematics and reading created by the current federal accountability system, noted above. The committee's proposed indicators to address this recommendation are directly related to the indicators on devoting more time and resources to science instruction and are designed to encourage state and federal policy makers to remedy the current imbalances.

Key Indicators to Monitor

Funding levels, accountability structures, and legislative mandates provide the most meaningful indicators of the value policy makers place on science education.

INDICATOR 9. Inclusion of science in federal and state accountability systems.

For each state, this indicator would measure whether science is assessed, whether science is included in the state accountability system, and whether science counts as much as reading and mathematics in that accountability system. Similarly, at the federal level, the indicator would capture whether or not the reauthorized Elementary and Secondary Education Act includes accountability provisions for science, and if so, in what grades.

Available and potentially available data. The Council of Chief State School Officers (CCSSO) has conducted surveys of states' accountability and assessment policies. Analyses of the survey results have reported whether science is assessed, but not whether science is part of the accountability system (Stillman and Blank, 2008). State applications for flexibility under the No Child Left Behind Act do indicate whether the states that have applied include science in their accountability systems: to date, less than half of them do.

Data and research needs. If the CCSSO survey cannot be modified to include questions about whether science is part of a state's accountability system, another organization that routinely conducts surveys of state educational policies, such as *Education Week* or the National Governors Association, might conduct a 50-state survey of the role of science in state accountability systems.

INDICATOR 10. Inclusion of science in major federal K-12 education initiatives.

Indicators of a federal commitment to science education could include the level of federal funding for the *Next Generation Science Standards* assessment relative to the *Common Core State Standards for English Language Arts and Mathematics*; the inclusion of science in major federal incentive programs such as Race to the Top; whether funding in support of science education is included in the reauthorization of the Elementary and Secondary Education Act; and the frequency of science assessment in NAEP, compared with reading and mathematics.

Available and potentially available data. Information about these indicators could be gathered by analyzing publicly available budget data; federal solicitations for the largest grant programs and other competitive funding; and eventually, language in the reauthorized Elementary and Secondary Education Act. Information about the frequency of NAEP assessments is well known and readily available from the NCES.

INDICATOR 11. State and district staff dedicated to supporting science instruction.

This indicator is intended to measure the human resources that are available for science relative to mathematics and reading for all grade levels. It should include the various types of human resources that are devoted to these subjects at the district and state levels, such as curriculum specialists, coaches, and teacher specialists. For example, as research identifies the expertise needed for high-quality teaching of specific content domains, elementary schools are increasingly moving away from the traditional model of one elementary teacher who teaches all four content areas and beginning to use specialists (e.g., Association of Mathematics Teacher Educators, 2010). These specialists are often district staff who serve multiple schools and take on a variety of roles, including teaching students, managing materials, planning with the classroom teachers, conducting professional development, and offering demonstration lessons (Century, Rudnick, and Freeman, 2008).

Available and potentially available data. No data are currently collected that provide complete information about this indicator on a large scale. The SASS includes questions about the percentage of schools with science and mathematics coaches and science specialists. The SASS district survey, the CCSSO 50-state survey, or *Education Week*'s Quality Counts survey could be used to collect district and state data, respectively. The Common Core of Data state survey includes questions about instructional coordinators and supervisors; these questions could be revised to be specific to reading, mathematics, and science. Relevant questions from a survey of science education in California (Dorph et al., 2011) also might be adapted for use on a national scale.

Data and research needs. Additional research is needed to learn whether elementary students' science learning opportunities are of high quality and whether their science learning is enhanced when they receive science instruction from a science specialist rather than a generalist teacher. Research on the most effective use of content-focused coaches (e.g., modeling lessons, coaching individual teachers, providing professional development to groups of teachers) also would be productive.

Effective Systems of Assessment

Indicator 10 is proposed to suggest that states and the federal government can help elevate science to the same level of importance as reading and mathematics by including more science assessment in accountability systems. Yet, as with reading and mathematics assessments, it is crucial that science assessments support, rather than undercut, effective science instruction. Thus, in addition to developing quality assessments to accompany the *Common Core State Standards for Mathematics,* it is necessary to develop science assessments that reflect current research on teaching and learning and that emphasize the practices of science before including science in accountability systems. Indeed, *Successful K-12 STEM Education* recommended that "states and national organizations should develop effective systems of assessment that ... emphasize science practices rather than mere factual recall" (National Research Council, 2011, p. 28).

Key Indicator to Monitor

The committee proposes one indicator related to state assessment systems. Collecting data for this indicator involves analyzing the mathematics and science assessment systems that states adopt in the coming years, to determine the extent to which they embody the *Common Core State Standards for Mathematics* and the vision for science education in *A Framework for K-12 Science Education* (National Research Council, 2012) or the extent to which those assessment systems otherwise emphasize mathematical and science practices in addition to concept mastery. In this way, it will be possible to identify states with assessment systems that support effective instruction.

We include mathematics in this indicator because the concern about quality assessments is not limited to science. If this indicator is to measure assessments that accompany new standards in mathematics and science, no data could be collected on it until the next generation of assessments has been developed and states have adopted them.

INDICATOR 12. States' use of assessments that measure the core concepts and practices of science and mathematics disciplines.

The committee proposes a two-tiered data collection for this indicator—similar to that proposed for Indicator 4—that involves determining which assessments states use and then analyzing the most commonly used assessments for their consistency with standards documents.

Available and potentially available data. The CCSSO has conducted a 50-state survey of state assessment polices, and their report includes the most commonly used assessments in science (Stillman and Blank, 2008). The survey should be administered again when states adopt assessments related to the *Common Core State Standards for Mathematics* and *A Framework for K-12 Science Education*. However, such a survey would only provide information on the first tier of data collection: which assessments states use.

Data and research needs. If the CCSSO 50-state survey cannot be readministered, similar surveys of state departments of education, analyses of state education websites, and other means could yield information about what assessments are being used by each state, as well as accommodations that are offered students. Other research would need to be conducted on the qualities of the assessments, including their rigor, reliability, and validity. Efforts to develop a procedure for analyzing the degree to which assessments embody the standards or emphasize the practices of science could build on similar efforts to evaluate curricular materials by the American Association for the Advancement of Science (2005); Schmidt et al. (1997); and Gueudet, Pepin, and Trouche (2012).

Federal and State Support for STEM Teachers

Successful K-12 STEM Education recommended that "National and state policy makers should invest in a coherent, focused, and sustained set of supports for STEM teachers to help them teach in effective ways" (National Research Council, 2011, p. 28). Although districts bear most of the responsibility for providing professional development for teachers, federal and state agencies provide the bulk of funding for these activities. Most states and districts do not develop and implement strategic plans to bolster science and mathematics teaching, which poses a significant hurdle to supporting teachers' continued growth and development throughout their careers (Borko, 2004; Wilson, Rozelle, and Mikeska, 2011). Many districts lack a mechanism for coordinating a focused portfolio of professional development that is aligned with instructional reforms. Some schools send single volunteers to a professional development program, with the understanding that a teacher who returns will share what she has learned with her colleagues. Other schools have professional development offered by the developers of new textbooks or instructional materials that they have adopted. In yet other schools, teachers participate in professional development through NSF projects at local universities or educational organizations.

A coherent strategy for investing in high-quality learning opportunities for science and mathematics teachers would represent a promising step toward enhancing and maintaining their capacity to teach (Cobb and Smith, 2008). Such a strategy might include crafting teacher induction programs that are aligned with later professional development so that teachers can deepen their understanding of STEM disciplines, STEM teaching, and how their students understand STEM concepts. It also might involve identifying the core concepts that teachers need to understand to be able to teach to the Common Core State Standards and the core ideas, practices, and crosscutting concepts in *A Framework for K-12 Science Education* (National Research Council, 2012) and developing a long-term professional development program that would gradually deepen teachers' knowledge and skill in teaching those topics.

Key Indicator to Monitor

Efforts to monitor progress toward this recommendation from *Successful K-12 STEM Education* should concentrate on the kinds of professional development activities that are supported by federal and state funding, in order to determine whether that funding is supporting coherent activities that are consistent with best practices identified by the research. In turn, these efforts should be used to generate information about the characteristics of professional development activities that lead to changes in teaching practice and to improved student outcomes.

INDICATOR 13. State and federal expenditures dedicated to improving the K-12 STEM teaching workforce.

This indicator should measure expenditures on preparation, recruitment, induction, recruitment, and subject-specific professional development over teachers' careers. Examples of programs whose expenditures might be tracked include UTeach, NSF's Mathematics and Science Partnerships, the California State University's Math and Science Teacher Initiative, and the National Aeronautics and Space Administration (NASA) Pre-service Teacher Institute. The indicator should enable distinction among different kinds of activities, to facilitate future analyses of the extent to which those activities are evidence based.

Available and potentially available data. The NCES does not collect data on this indicator in a consistent way. At the federal level, this information could be gathered by identifying and totaling the grants or grant programs that support teacher training and professional development at NSF, the U.S. Department of Education (Title II), and other agencies such as NASA and the U.S. Department of Energy. This approach could also be repeated for each state, though it would be considerably more complex and time consuming.

Data and research needs. New kinds of data collection would be needed to more systematically track district and state investments in recruitment, induction, and professional development. Addressing the research needs identified under Indicators 7 and 8 also would help to identify whether the spending on these activities is evidence based.

Research to Enhance Understanding of STEM Schools, Practices, and Outcomes

Successful K-12 STEM Education recommended that "federal funding for STEM-focused schools should be tied to a robust, strategic research agenda" (National Research Council, 2011, p. 28). That recommendation also noted that such an agenda would include research that

- disentangles the effects of school practice from student selection;
- takes into account the importance of contextual variables on teaching, learning, and student outcomes; and
- allows for longitudinal assessment of student outcomes, including the three strategic goals of U.S. education in STEM and intermediate outcomes relative to those goals.

These research needs can be met with a variety of methodologies. For example, it would be useful to have descriptive information on the number of people, schools, districts, and/or states offering or participating in a given activity or practice (e.g., the number of STEM-focused schools or programs, principals' participation in professional development) and the degree to which these activities are addressing the needs of students from traditionally low-performing populations. Contrast studies with comparison or control groups would allow for determinations of the effectiveness of instructional strategies or models, curricula, and professional development activities in improving student learning, engagement, and persistence. Interview studies, in-depth case studies, and classroom observations would be useful for understanding how practices and policies are being implemented, and for illuminating contextual influences on the teaching and learning process. And by blending theory building and the development of design principles, design-based research would guide, inform, and improve practice and research (Anderson and Shattuck, 2012). Research also is needed to develop appropriate instruments and to learn how to support implementation at scale of effective science and mathematics initiatives. Regardless of the design, this research should be conducted on a focused set of topics to generate information on the most pressing questions in science and mathematics education.

Key Indicator to Monitor

INDICATOR 14. Federal funding for the research identified in *Successful K-12 STEM Education*.

Available and potentially available data. The U.S. Department of Education's Institute for Education Sciences (IES), various NSF directorates, and the National Institutes of Health (NIH) are the primary funders of the kinds of research recommended in *Successful K-12 STEM Education*. Data on the types of research funded and the knowledge accumulated from those investments currently are not compiled in a way that would provide readily accessible information for this indicator.

Data and research needs. Information on this indicator could be gleaned by analyzing the portfolio of grants awarded by IES, NSF, and NIH. That analysis could examine how many studies, and what proportion, addressed the research gaps identified in *Successful K-12 STEM Education*.

Another approach would be to examine the pattern of publications in key journals, such as the *American Educational Research Journal*, *Educational Evaluation and Policy Analysis*, the *Journal for Research in Mathematics Education*, and the *Journal for Research in Science Teaching*.[6] That examination could determine how many studies address each of the gaps identified in *Successful K-12 STEM Education* and whether the pattern is changing over time.

[6]In 2003, IES conducted an analysis similar to the one proposed here. The presentation of that analysis is available at http://ies.ed.gov/director/speeches2003/04_22/2003_04_22b.asp [August 2012].

CREATING A MONITORING AND REPORTING SYSTEM FOR K-12 STEM EDUCATION

The committee's intent is for Indicators 1-14 to form the core of a national program to monitor the health of the education system by gathering information from the national, state, and local levels. Although these indicators were developed in conjunction with a specific set of recommendations from a previous report, when considered together they have the potential to provide insights into key elements of the K-12 education system in STEM that are difficult to glean from current data collection systems. However, additional research and data beyond the indicators would be required to undertake a full-scale evaluation of the nation's progress toward the recommendations in *Successful K-12 STEM Education,* and to link the inputs of the education system to outcomes.

First, the monitoring and reporting system would be more meaningful if it included measurements of progress toward the goals of increasing the number of underrepresented students who pursue science and engineering degrees and careers, expanding the STEM-capable workforce, and increasing science literacy, because those goals provide the context for the recommendations in *Successful K-12 STEM Education.* Although district and state data are valuable as system-level indicators, it would be possible to track progress toward the goals for STEM education and provide a more comprehensive portrait of education and students' experiences in STEM by measuring a wide range of student-level outcomes in the following categories:

- **K-12 academic achievement and participation in science and mathematics** (e.g., conceptual understanding; proficiency with the practices of science, engineering, and mathematics; science and mathematics course-taking patterns; enrollment in technical training programs while in high school).
- **Values, attitudes, and beliefs about STEM** (e.g., students' fascination with natural and physical phenomena, interest in and value of science, beliefs about their competence in science, identities as science learners).
- **Access to and participation in STEM-related activities** (e.g., authentic research experiences and internships; interactions with adult mentors; out-of-school time science activities such as science clubs and competitions).
- **Postsecondary training and education in the STEM disciplines** (e.g., intention to study STEM expressed in K-12; accumulation of college credit in STEM courses; degrees and certificates earned).
- **Participation in STEM-related careers** (e.g., intention to pursue STEM-related careers and STEM-related career counseling in K-12; eventual participation in STEM workforce).

Because access and participation are vital for students from groups that are underrepresented in STEM, it is important to track each of these outcomes by race, ethnicity, language status, and socioeconomic status.

Through various NCES surveys, NSF's *Science and Engineering Indicators,* and the Bureau of Labor Statistics, national-level data are widely available on certain aspects of the outcomes in these categories—albeit to a lesser extent for measures of students' values, attitudes, and beliefs. Further research would be needed to develop appropriate measures of affective indicators. Although previous efforts to develop such measures have faced challenges, promising efforts are currently under way (e.g., Hidi and Renninger, 2006; Dorph et al., 2012). In addition, some of the existing measures may require refinement to enable disaggregation of data for different student groups. They also might need refinement to reflect a broad definition of postsecondary education that includes technical training with or without certification, 2-year colleges, 4-year colleges, and postbaccalaureate study, and a broad definition of STEM-related careers that includes such occupations as health care and energy technicians and science and mathematics teachers, in addition to scientists, engineers, and mathematicians.

Second, the information that can be gleaned from any set of indicators is necessarily limited. The proposed indicators will yield data that enable counts, classification, and the tracking of trends on the national level. Other kinds of research are needed to explain those trends and to understand how the recommendations from *Successful K-12 STEM Education* are being implemented in various educational contexts. Balancing data from the indicators with richer, more localized sources of information such as case studies, interviews, and classroom observations would enable a more complete description of K-12 STEM education.

Characteristics and Capabilities

To generate relevant information that can be used for improvement, the monitoring system would be designed with the capability to:

- assess progress toward key improvements recommended in the 2011 National Research Council report *Successful K-12 STEM Education;*
- measure student knowledge, interest, and participation in the STEM disciplines and STEM-related activities;
- track financial, human capital, and material investments in K-12 STEM education at the federal, state, and local levels;
- provide information about the capabilities of the STEM-education workforce, including teachers and principals; and
- facilitate strategic planning for federal investments in STEM education and workforce development, when used with labor force projections.

By focusing on student subgroups and the nation as a whole, these capabilities could be met in a way that illuminates variations in access, opportunities to learn, and progress as they relate to different student populations and socioeconomic contexts.

Effective monitoring and reporting systems are dynamic and evolve over time. The committee developed its proposed indicators in the context of the adoption and implementation of the *Common Core State Standards for Mathematics* and the development, adoption, and eventual implementation of the *Next Generation Science Standards* based on *A Framework for K-12 Science Education* (National Research Council, 2012). Both developments will be driving U.S. mathematics and science education for the foreseeable future. Different indicators might be warranted as these reforms mature, and as additional research becomes available on such factors as school conditions, teacher quality, effective instructional practices, and student performance measures. As this research emerges along with data from the indicators, it will be important to periodically reassess the continuing relevance and utility of the data that are generated, and to refine and adjust the indicators as necessary.

The committee's intent is for the proposed monitoring and reporting system to monitor progress in K-12 STEM education and promote the organizational practices of examining data and using the results of such examination to take the necessary actions to improve performance and attain the desired goals (U.S. Department of Education, 2010). Thus, the monitoring system should be designed to promote the practices of continuous quality improvement toward achieving the nation's goals for education and workforce development in STEM.

Plan for Implementation

The range of activities described below can be launched immediately to begin developing the full suite of proposed indicators. Many of these activities can be undertaken simultaneously. However, because it may not be possible to undertake work on all aspects of the proposed 14 indicators immediately, the committee identified 6 of them as the highest priorities (see Box 2).

Identify Data Collection Mechanisms

For each of the 14 indicators presented in the preceding section, we identified available and potentially available sources of data and discussed some limitations of those data sources as they relate to specific indicators. On the whole, existing data systems do not currently provide all of the needed information for the proposed indicators (see Table 1). Currently, the full complement of data is only available for Indicators 10, 13, and 14. This scarcity of data reflects the fact that existing federal data collection systems primarily provide information about *schools* rather than *schooling* (the process by which students learn). Many national surveys ask about such characteristics as the number of teachers in a school in each subject, whether the school has a theme, and the number of students taking advanced courses.

BOX 2
Priority Indicators

The committee's intent is for efforts to be undertaken now to establish a system for collecting information on all 14 indicators. However, if resources for the monitoring system are too limited to support full implementation, the committee has identified six indicators as being of the highest priority:

- **Indicator 2.** Time allocated to teach science in grades K-5.
- **Indicator 4.** Adoption of instructional materials in grades K-12 that embody the *Common Core State Standards for Mathematics* and *A Framework for K-12 Science Education*.
- **Indicator 5.** Classroom coverage of content and practices in the *Common Core State Standards for Mathematics* and *A Framework for K-12 Science Education*.
- **Indicator 6.** Teachers' science and mathematics content knowledge for teaching.
- **Indicator 9.** Inclusion of science in federal and state accountability systems.
- **Indicator 14.** Federal funding for the research identified in *Successful K-12 STEM Education*.

The committee selected these six indicators in the belief that they represent the points of greatest leverage to improve the education system, student outcomes in the STEM disciplines, and progress toward the three goals of education in STEM. The first five priority indicators (2, 4, 5, 6, and 9) reflect conditions that are at the core of teaching and learning: time, materials, instruction, teacher knowledge, and accountability. The sixth priority indicator (14) calls for new research to fill critical gaps in knowledge about programs and practices that contribute to student learning and to the other goals of STEM education.

Of the priority indicators, 4, 5, and 6 will be the most resource intensive to develop, in part because they have never before been tracked on a large scale and existing measures do not take into account the current emphasis on the practices of science, mathematics, and engineering. Crafting a valid, independent procedure for analyzing instructional materials will constitute the bulk of the effort for Indicator 4. For Indicators 5 and 6, some relevant measures exist of classroom coverage and teachers' content knowledge for teaching—mostly in mathematics. These measures have not been taken to scale and further work would be required to develop direct measures of teachers' knowledge of the practices of science, mathematics, and engineering.

The effort required to develop these priority indicators is directly related to the current lack of useful data to understand and inform decision making about the issues that matter most to teaching and learning. Despite the effort involved, the committee deemed these indicators as high priority because at the dawn of new reforms in science, mathematics, engineering, and technology education, there is great need for sound data on these phenomena.

In contrast, the committee's proposed indicators address core elements of the learning process: the content and quality of the curriculum, the opportunities students have to learn that content, and teachers' knowledge for teaching science, mathematics, and engineering. This focus represents a significant shift in emphasis for state and federal data systems. This shift in focus presents a dilemma about whether to use existing data sources or to create a new data collection vehicle dedicated to the proposed indicators.

Using existing data sources has advantages: existing data or reasonable modifications to scheduled national surveys could yield initial information for all 14 indicators; it is less expensive; some baseline data already exist; and the sampling frames enable comparisons across different subject areas. There are also disadvantages to relying on existing data sources: no single survey or cluster of surveys provides all of the needed information, a staggered rotation schedule for the surveys means that a full complement of current data is never available at any given time, considerable work would be required to modify the surveys in ways that would yield the desired information, and the varied sampling frames affect the ability to draw inferences across different levels of the education system. Moreover, relying on external data sources means ceding control over the focus of the surveys and the questions used. Changes by the survey organization may mean that some variables are no longer available, or would be based on a different implementation of the questionnaire. Even seemingly small changes to the response format of a question may have profound consequences and change the results in unexpected ways.

Further Develop the Proposed Indicators

This report presents a framework for developing an indicator system around key elements of K-12 education in STEM. Although it was beyond the scope of the study to provide detailed specifications of the ways in which these indicators should be implemented, the committee assumes that the implementers will undertake a rigorous process to more fully develop a set of valid and reliable indicators with strong associations to the desired outcomes. For each indicator, the development process would include defining the construct, identifying what is known about what constitutes quality (i.e., what predicts downstream impact), identifying the most appropriate sources of data for broad measurement of the indicator, and identifying the most important topics to study in more depth than is possible with large-scale surveys. As one example, the committee's initial development of Indicators 1, 2, 3, 5, and 8 emphasizes quantity or prevalence; fully developed versions of these indicators also would measure quality. In most of those cases, additional research will be required to identify what constitutes quality and how to measure it.

The state of development of the proposed indicators varies, and the committee created the following four categories to classify their relative states of development:

TABLE 1 Summary of Indicators

	Recommendations from *Successful K-12 STEM Education*	**Indicators**	**State of Development**			
			Type			
			1	2	3	4
Access to Quality STEM Learning	*Districts Should Consider All Three Models of STEM-Focused Schools*	1. Number of, and enrollment in, different types of STEM schools and programs in each district.[a,b]		✓		✓
	Districts Should Devote Adequate Instructional Time and Resources to Science in Grades K-5	2. Time allocated to teach science in grades K-5.[a,b]	✓	✓		
		3. Science-related learning opportunities in elementary schools.[a,b]		✓		
	Districts Should Ensure That Their STEM Curricula Are Focused on the Most Important Topics in Each Discipline, Are Rigorous, and Are Articulated as a Sequence of Topics and Performances	4. Adoption of instructional materials in grades K-12 that embody the *Common Core State Standards for Mathematics* and *A Framework for K-12 Science Education.*[a]		✓		✓
		5. Classroom coverage of content and practices in the *Common Core State Standards for Mathematics* and *A Framework for K-12 Science Education.*[a,b]	✓	✓	✓	
Educators' Capacity	*Districts Need to Enhance the Capacity of K-12 Teachers*	6. Teachers' science and mathematics content knowledge for teaching.[a]	✓	✓	✓	✓
		7. Teachers' participation in STEM-specific professional development activities.[a]	✓	✓	✓	✓
	Districts Should Provide Instructional Leaders with Professional Development That Helps Them to Create the School Conditions That Appear to Support Student Achievement	8. Instructional leaders' participation in professional development on creating conditions that support STEM learning.[a,b]		✓	✓	✓
Policy and Funding Initiatives	*Policy Makers at the National, State, and Local Levels Should Elevate Science to the Same Level of Importance as Reading and Mathematics*	9. Inclusion of science in federal and state accountability systems.	✓	✓		
		10. Inclusion of science in major federal K-12 education initiatives.	✓			
		11. State and district staff dedicated to supporting science instruction.	✓	✓		
	States and National Organizations Should Develop Effective Systems of Assessment That Are Aligned with A Framework for K-12 Science Education *and That Emphasize Science Practices Rather Than Mere Factual Recall*	12. States' use of assessments that measure the core concepts and practices of science and mathematics disciplines.	✓	✓		✓
	National and State Policy Makers Should Invest in a Coherent, Focused, and Sustained Set of Supports for STEM Teachers	13. State and federal expenditures dedicated to improving the K-12 STEM teaching workforce.	✓			
	Federal Agencies Should Support Research That Disentangles the Effects of School Practice from Student Selection, Recognizes the Importance of Contextual Variables, and Allows for Longitudinal Assessments of Student Outcomes	14. Federal funding for the research identified in *Successful K-12 STEM Education.*	✓			

- Type 1: At least some data currently are available through U.S. Department of Education surveys or other large-scale efforts. Ongoing development may be required to more fully develop the indicator.
- Type 2: Appropriate data can be collected by modifying existing U.S. Department of Education surveys. Conceptual or empirical work may be required to develop valid and reliable survey items.
- Type 3: New surveys might be required to collect appropriate data, which would involve conceptual and empirical development.
- Type 4: Conceptual and empirical development are required to begin specifying the indicator.

Each type corresponds to the level of resources required to fully develop the indicator, which increases from 1 to 4. For example, it is more costly and time consuming to fund a systematic research and development effort than it is to compile available data. Similarly, modifying questions on existing surveys is less resource intensive than creating entirely new surveys.

For Type 1 indicators that fall into other categories in Table 1, existing data might provide partially useful information, but revising existing surveys would yield more relevant information about the indicator in question. In those cases, conceptual and empirical work are required to develop survey items or more fully specify the indicator as initial data are collected. For indicators that span all four types, modifications to existing surveys might be useful in the short term; but in the longer term, other kinds of data collection might be more appropriate, and more work is required to fully develop the indicator.

Compile, Analyze, and Report on Existing Data for Type 1 Indicators

As discussed above, existing national data systems provide at least partial coverage for the Type 1 indicators. Through grant awards, agency budgets, legislative language, and state applications for flexibility under the No Child Left Behind Act, there is information on several other indicators, but the relevant documents would have to be gathered and analyzed to yield the data related to the indicators. The committee estimates that within 1 to 2 years, NCES and NSF could compile and analyze relevant data that have already been collected and use those analyses to produce an initial status report. Such a report might include a discussion of all available data on the Type 1 indicators (2, 5, 6, 7, 9, 10, 13, and 14) along with presentations of student-level data that illustrate the nation's status relative to the three goals for education in STEM. An analysis of gaps in the available data might lead to conclusions and recommendations about developing the full suite of indicators, which could be discussed in the report.

TABLE NOTES: Shaded areas represent the committee's highest priorities. These indicators are most proximal to the core of teaching and learning.

[a]Data should be disaggregated to report on different groups of students and to facilitate analyses of how the indicators vary with the socioeconomic status of states or school districts.

[b]Initial development of this indicator emphasizes quantity; full development also should include quality.

Modify Existing Surveys or Develop New Data Collection Mechanisms for Type 2 and 3 Indicators

If the monitoring system relies on existing data sources, revising existing or past questions or adding new questions to existing NCES surveys could yield the needed information on some of the Type 2 indicators (notably, 2, 3, 9, and 11). In those cases, NCES could begin the modifications immediately so that they are complete by the next data collection for the relevant surveys: see the Appendix, which shows that the next data collection for most relevant surveys ranges from the 2012-2013 academic year to the 2015-2016 academic year.

Development of items for a dedicated survey or surveys also could begin immediately. Regardless of which data collection mechanism is chosen, careful attention should be given to developing and validating survey items so that they provide reliable measures of the specified indicator.

Support Work That Is Needed to Further Develop Type 4 Indicators

For the Type 4 indicators (1, 4, 6, 7, 8, and 12), research or conceptual development is required to identify what the indicator should measure and how it should be measured. For example, Indicators 4 and 12 require the development of new procedures to analyze the alignment of instructional materials and assessments with standards documents; the understanding of those indicators will remain limited until such procedures are developed. In a different vein, general information currently could be collected about instructional leaders' participation in professional development (Indicator 8). However, the research bases on what constitutes effective leadership for STEM and on which kinds of professional development support that leadership are not yet robust enough to identify exactly what this indicator should measure. Further research on those topics is needed to modify existing surveys in a way that will generate useful information.

IES and NSF could support much of the needed research and development through existing grant programs. The research community also might be enticed to pursue this work through new solicitations or Dear Colleague letters. Given the steps involved to solicit, fund, and conduct the needed research, the committee estimates that acquisition of the full suite of data on these indicators would take approximately 5-10 years from the time the effort is launched.

Produce Regular Reports on the Indicators and STEM Education Goals

Despite the availability of some data that are relevant to the committee's proposed indicators, these data are not regularly analyzed and compiled in a report that is focused on K-12 education in the STEM disciplines. As the summary in the Appendix shows, most NCES survey data are not regularly compiled for publication in reports. Instead, data tables from many of the surveys are publicly available. NCES annual reports (*The Condition of Education, Digest of Education Statistics, High School Dropout and Completion Rates,* and *Indicators of School Crime and Safety*) typically are broad in focus. Some spinoff reports from the *Condition of Education* have addressed the STEM disciplines (e.g., National

Center for Education Statistics, 1997a, 1997b), and NSF's annual *Science and Engineering Indicators* report includes a chapter on K-12 science and mathematics education. However, NCES and NSF do not regularly publish reports focused solely on K-12 education in the STEM disciplines. In the 1990s, NSF produced two biennial reports on K-12 science and mathematics education indicators apart from the *Science and Engineering Indicators,* but that reporting program was discontinued in 1995 (for the last of these reports, see National Science Foundation, 1996).

It would be valuable for NCES or NSF to use data from the proposed monitoring system to begin producing a biennial report that is unique and specific to the issue of K-12 education in the STEM disciplines (along the lines of reports requested by Congress on dropouts and the implementation of the Individuals with Disabilities Education Act). This report could present data that demonstrate the nation's progress with respect to the proposed indicators and the goals for K-12 education in STEM. Consistent with the proposed capabilities of the monitoring system, the report might analyze the currently available data to provide usable information on student knowledge, interest, and participation in the STEM disciplines and STEM-related activities; federal, state, and local investments in K-12 education in STEM; and the capabilities of the STEM-education workforce.

Support Efforts to Engage Stakeholders

The U.S. Department of Education and NSF could play a key role in developing and sustaining the proposed monitoring system by engaging stakeholders in ongoing discussions about inputs and outputs to the system. Such efforts could increase the understanding of the monitoring system; inform the development, definition, and refinement of the indicators; and ensure that the data are used to support improvements. As an example, periodic conferences linked to the reports described above could address the nation's progress toward the nation's STEM education goals and the recommendations in *Successful K-12 STEM Education.* Such conferences could engage education leaders, policy makers, researchers, and science and engineering professional societies in general discussions of the current state of K-12 education in STEM; in focused discussions of data related to specific indicators; in proposed adjustments to the education system in light of the nation's progress; and in critical examinations of the ongoing utility of the indicators.

CONCLUSION

The committee's proposed indicator system offers what we think is an important new way of advancing understanding of the state of STEM education and the development of the STEM workforce by meaningfully addressing the complex processes of schooling. It will also enable measuring progress toward the nation's goals in these critical arenas. The proposed system represents a significant departure from existing data collection systems by linking both inputs and outcomes, and by moving beyond measures of academic achievement to recognize that variables related to student engagement and life choices also are important to meeting the goals for U.S. K-12 education in STEM.

The time to put this monitoring and reporting system into place could not be more opportune. In this era of heightened accountability in education, the availability of and capacity to collect high-quality data are greater than ever before. Moreover, with the advent of new standards in mathematics and science, the demand is increasing for data that measure the key elements of those standards. An exceptional opportunity exists to collect baseline data as states and districts begin implementing the new standards in the coming years. The committee's proposed indicators are designed to capitalize on current opportunities and make a meaningful contribution to ongoing efforts to improve K-12 education in STEM by providing needed data to make informed decisions.

REFERENCES

American Association for the Advancement of Science. (2005). *High school biology textbooks: A benchmarks-based evaluation*. Washington, DC: American Association for the Advancement of Science. Available: http://www.project2061.org/publications/textbook/hsbio/report/default.htm [July 2012].

Anderson, T., and Shattuck, J. (2012). Design-based research: A decade of progress in education research? *Educational Researcher, 41*(1), 16-25.

Association of Mathematics Teacher Educators. (2010). *Standards for elementary mathematics specialists: A reference for teacher credentialing and degree programs*. San Diego, CA: Association of Mathematics Teacher Educators.

Banilower, E., Boyd, S.E., Pasley, J.D., and Weiss, I.R. (2006). *Lessons from a decade of mathematics and science reform: A capstone report for the Local Systemic Change through Teacher Enhancement Initiative.* Chapel Hill, NC: Horizon Research. Available: http://www.horizon-research.com/pdmathsci/htdocs/reports/capstone.pdf [October 2012].

Borko, H. (2004). Professional development and teacher learning: Mapping the terrain. *Educational Researcher, 33*(8), 3-15.

Bryk, A.S., and Driscoll, M.E. (1988). *The high school as community: Contextual influences and consequences for students and teachers*. Madison: National Center on Effective Secondary Schools, University of Wisconsin.

Bryk, A.S., Sebring, P.B., Allensworth, E., Luppescu, S., and Easton, J.Q. (2010). *Organizing schools for improvement: Lessons from Chicago*. Chicago, IL: University of Chicago Press.

Carnevale, A.P., Smith, N., and Melton, M. (2011). *STEM*. Washington, DC: Georgetown University Center on Education and the Workforce. Available: cew.georgetown.edu/STEM [August 2012].

Center on Education Policy. (2007). *Choices, changes, and challenges: Curriculum and instruction in the NCLB era*. Washington, DC: Center on Education Policy.

Century, J., Rudnick, M., and Freeman, C. (2008). Accumulating knowledge on elementary science specialists: A strategy for building conceptual clarity and sharing findings. *Science Educator, 17*(2), 31-44.

Cobb, P., and Smith, T.M. (2008). The challenge of scale: Designing schools and districts as learning organizations for instructional improvement in mathematics. In K. Krainer and T. Wood (Eds.), *International handbook of mathematics teacher education: Vol. 3. Participants in mathematics teacher education: Individuals, teams, communities and networks* (pp. 231-254). Rotterdam, The Netherlands: Sense.

Cobb, P., Jackson, K., Smith, T.M., Sorum, M., and Henrick, E. (in press). Design research within educational systems: Investigating and supporting improvements in the quality of mathematics teaching and learning at scale. In W. Penuel, B. Fishman, and B.H. Cheng (Eds.), *Design-based implementation research.* National Society for the Study of Education (NSSE) Yearbook. Available: http://peabody.vanderbilt.edu/docs/pdf/tl/NSSE_Yearbook_Chpt_2nd_Draft_120816.pdf [January 2013].

DeAngelis, K.J., and Presley, J.B. (2011). Teacher qualifications and school climate: Examining their interrelationship for school improvement. *Leadership and Policy in Schools, 10*(1), 84-120.

Desimone, L. (2009). How can we best measure a teacher's professional development and its effects on teachers and students? *Educational Researcher, 38*(3), 181-199.

Dorph, R., Shields, P., Tiffany-Morales, J., Hartry, A., and McCaffrey, T. (2011). *High hopes—few opportunities: The status of elementary science education in California*. Sacramento, CA: The Center for the Future of Teaching and Learning at WestEd. Available: http://www.lawrencehallofscience.org/sites/lawrencehallofscience.org/files/user-jnoe/ScienceFullReportweb.pdf [July 2012].

Dorph, R., Schunn, C., Crowley, K., and Shields, P. (2012). *Activating young science learners: Igniting persistent engagement in science learning and inquiry*. A structured poster session presented at the American Education Research Association Annual Meeting, Vancouver, British Columbia.

Gamoran, A., Anderson, C.W., Quiroz, P.A., Secada, W.G., Williams, T., and Ashmann, S. (2003). *Transforming teaching in math and science: How schools and districts can support change*. New York: Teachers College Press.

Garet, M.S., Porter, A.C., Desimone, L., Birman, B.F., and Yoon, K.S. (2001). What makes professional development effective? Results from a national sample of teachers. *American Educational Research Journal, 38*(4), 915-945.

Garet, M.S., Wayne, A., Stancavage, F., Taylor, J., Eaton, M., Walters, K., Song, M., Brown, S., Hurlburt, S., Zhu, P., Sepanik, S., and Doolittle, F. (2011). *Middle school mathematics professional development impact study: Findings after the second year of implementation*. NCEE 2011-4024. Washington, DC: National Center for Education Evaluation and Regional Assistance, Institute of Education Sciences, U.S. Department of Education.

Gueudet, G., Pepin, B., and Trouche, L. (Eds.). (2012). *Mathematics curriculum material and teacher development: From text to "lived" resources*. Dordrecht, The Netherlands: Springer.

Hartry, A., Dorph, R., Shields, P., Tiffany-Morales, J., and Romero, V. (2012). *The status of middle school science education in California*. Sacramento, CA: The Center for the Future of Teaching and Learning at WestEd. Available: http://www.lawrencehallofscience.org/sites/lawrencehallofscience.org/files/user-jnoe/Middle_School_Science_Ed_%20Study.pdf [October 2012].

Hidi, S., and Renninger, K.A. (2006). The four-phase model of interest development. *Educational Psychologist, 41*(2), 111-127.

Higher Education Research Institute. (2010). *Degrees of success: Bachelor's degree completion rates among initial STEM majors*. Available: http://www.heri.ucla.edu/nih/downloads/2010%20-%20Hurtado,%20Eagan,%20Chang%20-%20Degrees%20of%20Success.pdf [August 2012].

Hill, H.C., and Ball, D.L. (2004). Learning mathematics for teaching: Results from California's Mathematics Professional Development Institutes. *Journal of Research in Mathematics Education,35*(5), 330-351.

Hill, H.C., Schilling, S.G., and Ball, D.L. (2004). Developing measures of teachers' mathematics knowledge for teaching. *Elementary School Journal, 105*(1), 11-30.

Hill, H.C., Rowan, B., and Ball, D.L. (2005). Effects of teachers' mathematical knowledge for teaching on student achievement. *American Educational Research Journal, 42*(2), 371-406.

Hill, C.J., Bloom, H.S., Black, A.R., and Lipsey, M.W. (2008). Empirical benchmarks for interpreting effect sizes in research. *Child Development Perspectives, 2*(3), 172-177.

Hochberg, E., and Desimone, L. (2010). Professional development in the accountability context: Building capacity to achieve standards. *Educational Psychologist, 45*(2), 89-106.

Hudson, S.B., McMahon, K.C., and Overstreet, C.M. (2002). *The national survey of science and mathematics education: Compendium of tables*. Chapel Hill, NC: Horizon Research. Available: http://2000survey.horizon-research.com/reports/tables/tables_complete.pdf [October 2012].

Lowell, B.L., Salzman, H., Bernstein, H., and Henderson, E. (2009, November 5-7). *Steady as she goes? Three generations of students through the science and engineering pipeline.* Paper presented at Annual Meeting of the Association for Public Policy Analysis and Management, Washington, DC.

McLaughlin, M.W., and Talbert, J.E. (2006). *Building school-based teacher learning communities.* New York: Teachers College Press.

National Academy of Engineering. (2010). *Standards for K-12 engineering education?* Engineering Education Committee on Standards for K-12 Engineering Education. Washington, DC: The National Academies Press.

National Center for Education Statistics. (1997a). *Science proficiency and course taking in high school: The relationship of science course-taking patterns to increases in science proficiency between 8th and 12th grades.* NCES 97-838. Washington, DC: U.S. Department of Education.

National Center for Education Statistics. (1997b). *Women in mathematics and science: Findings from the condition of education 1997.* NCES 97-982. Washington, DC: U.S. Department of Education.

National Governors Association Center for Best Practices and Council of Chief State School Officers. (2010). *Common core state standards for mathematics.* Washington, DC: National Governors Association Center for Best Practices and Council of Chief State School Officers.

National Research Center for Career and Technical Education Curriculum Integration Workgroup. (2010). *Capitalizing on context: Curriculum integration in career and technical education.* Louisville, KY: National Research Center for Career and Technical Education.

National Research Council. (1988). *Improving indicators of the quality of science and mathematics education in grades K-12.* Committee on Indicators of Precollege Mathematics and Science Education. R.J. Murnane and S.A. Raizen (Eds.). Commission on Behavioral and Social Sciences and Education. Washington, DC: National Academy Press.

National Research Council. (1996). *National science education standards.* National Committee for Science Education Standards and Assessment. Washington, DC: National Academy Press.

National Research Council. (1999). *How people learn: Brain, mind, experience, and school.* Committee on Developments in the Science of Learning. J.D. Bransford, A.L. Brown, and R.R. Cocking (Eds.). Washington, DC: National Academy Press.

National Research Council. (2001). *Adding it up: Helping children learn mathematics.* Mathematics Learning Study Committee. J. Kilpatrick, J. Swafford, and B. Findell (Eds.). Center for Education. Division of Behavioral and Social Sciences and Education. Washington, DC: National Academy Press.

National Research Council. (2006). *America's lab report: Investigations in high school science.* Committee on High School Science Laboratories: Role and Vision. S.R. Singer, M.L. Hilton, and H.A. Schweingruber (Eds.). Board on Science Education, Center for Education. Division of Behavioral and Social Sciences and Education. Washington, DC: The National Academies Press.

National Research Council. (2007). *Taking science to school: Learning and teaching science in grades K-8.* Committee on Science Learning, Kindergarten Through Eighth Grade. R.A. Duschl, H.A. Schweingruber, and A.W. Shouse (Eds.). Board on Science Education, Center for Education. Division of Behavioral and Social Sciences and Education. Washington, DC: The National Academies Press.

National Research Council. (2009). *Learning science in informal environments: People, places, and pursuits.* Committee on Learning Science in Informal Environments. P. Bell, B. Lewenstein, A.W. Shouse, and M.A. Feder (Eds.). Board on Science Education, Center for Education. Division of Behavioral and Social Sciences and Education. Washington, DC: The National Academies Press.

National Research Council. (2010). *Preparing teachers: Building evidence for sound policy.* Committee on the Study of Teacher Preparation Programs in the United States, Center for Education. Division of Behavioral and Social Sciences and Education. Washington, DC: The National Academies Press.

National Research Council. (2011). *Successful K-12 STEM education: Identifying effective approaches in science, technology, engineering, and mathematics.* Committee on Highly Successful Science Programs for K-12 Science Education. Board on Science Education and Board on Testing and Assessment. Division of Behavioral and Social Sciences and Education. Washington, DC: The National Academies Press.

National Research Council. (2012). *A framework for K-12 science education: Practices, crosscutting concepts, and core ideas.* Committee on a Conceptual Framework for New K-12 Science Education Standards, Board on Science Education. Division of Behavioral and Social Sciences and Education. Washington, DC: The National Academies Press.

National Science Foundation Directorate for Education and Human Resources. (1996). *Indicators of science and mathematics education 1995.* NSF 96-52. L. Suter (Ed.). Arlington, VA: National Science Foundation.

OECD. (2012). *PISA in focus 18: Are students more engaged when schools offer extracurricular activities?* Paris: OECD.

Porter, A.C., Smithson, J., Blank, R., and Zeidner, T. (2007). Alignment as a teacher variable. *Applied Measurement in Education, 20*(1), 27-51.

Porter, A.C., Polikoff, M., and Smithson, J. (2009). Is there a de facto intended national curriculum? Evidence from state standards. *Educational Evaluation and Policy Analysis, 31*(3), 238-268.

President's Council of Advisors on Science and Technology. (2010). *Prepare and inspire: K-12 education in science, technology, engineering, and math (STEM) for America's future.* Washington, DC: President's Council of Advisors on Science and Technology. Available: http://www.whitehouse.gov/sites/default/files/microsites/ostp/pcast-stem-ed-final.pdf [July 2012].

PROM/SE. (2006). Knowing mathematics: What we can learn from teachers. *PROM/SE Research Report Series, 2.* Available: http://www.promse.msu.edu/_documents/PROMSE_KnowingMath.pdf [July 2012].

Schmidt, W.H. (2011, May). *STEM reform: Which way to go?* Paper presented at the National Research Council Workshop on Highly Successful STEM Education in K-12 Schools. Available: http://www7.nationalacademies.org/bose/STEM_Schools_Schmidt_Paper_May2011.pdf [August 2012].

Schmidt, W.H., and Maier, A. (2009). Opportunity to learn. In G. Sykes, B. Schneider, and N. Plank (Eds.), *Handbook of educational policy research* (pp. 541-560). New York and London: Routledge for the American Educational Research Association.

Schmidt, W.H., and McKnight, C.C. (2012). *Inequality for all.* New York: Teachers College Press.

Schmidt, W.H., McKnight, C.C., and Raizen, S.A. (1996). *Splintered vision: An investigation of U.S. science and mathematics education: Executive summary*. East Lansing: Michigan State University, U.S. National Research Center for the Third International Mathematics and Science Study.

Schmidt, W.H., McKnight, C.C., Valverde, G.A., Houang, R.T., and Wiley, D.E. (1997). *Many visions, many aims*. Dordrecht, The Netherlands: Kluwer Academic.

Schmidt, W.H., McKnight, C.C., Houang, R.T., Wang, H., Wiley, D.E., Cogan, L.S., and Wolfe, R.G. (2001). *Why schools matter: A cross-national comparison of curriculum and learning*. San Francisco, CA: Jossey-Bass.

Shadish, W., Cook, T.D., and Campbell, D.T. (2001). *Experimental and quasi-experimental designs for generalized causal inference*. Boston, MA: Houghton Mifflin.

Shavelson, R.J., McDonnell, L.M., and Oakes, J. (Eds.). (1989). *Indicators for monitoring mathematics and science education: A sourcebook*. Washington, DC: RAND Corporation. Available: http://www.rand.org/content/dam/rand/pubs/reports/2008/R3742.pdf [July 2012].

Silverberg, M., Warner, E.M., Fong, M., and Goodwin, D. (2004). *National assessment of vocational education: Final report to Congress*. Washington, DC: U.S. Government Printing Office.

Smith, P.S., and Taylor, M.J. (2010). *New tools for investigating the relationship between teacher content knowledge and student learning*. Presented at the 2010 NARST Annual Conference, Philadelphia, PA. Available: http://www.horizon-research.com/atlast/?page_id=103 [October 2012].

Smith, P.S., Banilower, E.R., McMahon, K.C., and Weiss, I.R. (2002). *The national survey of science and mathematics education: Trends from 1977-2000*. Chapel Hill, NC: Horizon Research. Available: 2000survey.horizon-research.com [October 2012].

Smithson, J., and Blank, R. (2006). *Indicators of quality teacher professional development and instructional change using data from surveys of enacted curriculum: Findings from NSF MSP-RETA Project*. Washington, DC: Council of Chief State School Officers.

Stein, M.K., and Nelson, B.S. (2003). Leadership content knowledge. *Educational Evaluation and Policy Analysis, 25*(4), 423-448.

Stillman, L., and Blank, R.K. (2008). *Key state education policies on PK-12 education: 2008*. Washington, DC: Council of Chief State School Officers. Available: http://www.ccsso.org/Documents/2008/Key_State_Education_Policies_2008.pdf [August 2012].

Stone, J.R., III, Alfeld, C., and Pearson, D. (2008). Rigor and relevance: Testing a model of enhanced math learning in career and technical education. *American Education Research Journal, 45*(3), 767-795.

Tai, R.H., Liu, C.Q., Maltese, A.V., and Fan, X. (2006). Planning early for careers in science. *Science, 312*(5777), 1143-1144.

Tatto, M.T., Schwille, J., Senk, S.L., Ingvarson, L., Peck, R., and Rowley, G. (2008). *Teacher Education and Development Study in Mathematics (TEDS-M): Policy, practice, and readiness to teach primary and secondary mathematics. Conceptual framework*. Amsterdam, The Netherlands: International Association for the Evaluation of Educational Achievement.

U.S. Department of Education. (2010). *Use of education data at the local level: From accountability to instructional improvement*. Washington, DC: U.S. Department of Education.

U.S. General Accounting Office. (2003). *Program evaluation: An evaluation culture and collaborative partnerships build agency capacity*. GAO-03-454. Washington, DC: U.S. Government Printing Office. Available: http://www.gao.gov/assets/240/238121.pdf [August 2012].

Valverde, G.A., Bianchi, L.J., Schmidt, W.H., McKnight, C.C., and Wolfe, R.G. (2002). *According to the book: Using TIMSS to investigate the translation of policy into practice in the world of textbooks*. Dordrecht, The Netherlands: Kluwer Academic.

Walpole, M., and Noeth, R.J. (2002). *The promise of Baldrige for K-12 education.* ACT Policy Report. Washington, DC: ACT.

Wilson, S.M., Floden, R.F., and Ferrini-Mundy, J. (2001). *Teacher preparation research: Current knowledge, recommendations, and priorities for the future*. Center for the Study of Teaching Policy, University of Washington, Seattle, WA.

Wilson, S.M., Rozelle, J., and Mikeska, J.N. (2011). Cacophony or embarrassment of riches: Building a system of support for teacher quality. *Journal of Teacher Education, 62*, 383-394.

Yoon, K.S., Duncan, T., Lee, S. W-Y., Scarloss, B., and Shapley, K. (2007). *Reviewing the evidence on how teacher professional development affects student achievement.* Issues & Answers Report, No. 033. Washington, DC: U.S. Department of Education, Institute of Education Sciences, National Center for Education Evaluation and Regional Assistance, Regional Educational Laboratory Southwest.

APPENDIX

SUMMARY OF RELEVANT SURVEYS ADMINISTERED BY THE U.S. DEPARTMENT OF EDUCATION

Survey and URL	Description/ Focus	Respondents	Frequency of Administration	How Data Are Compiled	Most Recent Data Collection	Next Expected Data Collection
Baccalaureate and Beyond Longitudinal Study (B&B:2008) http://nces.ed.gov/surveys/b&b/	Students' education and work experiences after they complete a bachelor's degree; special emphasis on the experiences of new elementary and secondary teachers	Nationally representative cohort of students who obtained a bachelor's degree in 2008	Cohort surveyed in 2009 (1st year out of college) and in 2012 (3rd year out of college)	Data tables available; too early for in-depth analysis	2012	Unknown
Beginning Postsecondary Students Longitudinal Study (BPS:04) http://nces.ed.gov/surveys/bps/	Student persistence in postsecondary education programs, their transition to employment, and demographic characteristics	Nationally representative cohort of students who entered postsecondary education in 2004	Cohort surveyed in 2004 (at the end of their 1st academic year), in 2006 (end of their 3rd academic year), and 2009 (end of their 6th academic year)	Data tables available; occasional use in analysis of community colleges (including 2009, 2008, 2007)	2009	Unknown
Civil Rights Data Collection (CRDC) http://ocrdata.ed.gov/	Information relating to providing equal educational opportunity, such as student enrollment and educational programs and services available	Survey of 100 percent of schools in a representative sample of districts across the country	Every 2 years	Data tables available; primarily used by the Office of Civil Rights for enforcement efforts	2011-2012	2013-2014
Common Core of Data (CCD) http://nces.ed.gov/ccd/	General descriptive information on schools and school districts, data on students and staff, and fiscal data	State education agencies respond from their own records	Annually	Used annually in reports of state and district finances, teacher compensation, and high school completion	2011-2012 (most recent available for release 2008-2009)	2012-2013

Survey and URL	Description/ Focus	Respondents	Frequency of Administration	How Data Are Compiled	Most Recent Data Collection	Next Expected Data Collection
Early Childhood Longitudinal Study-Kindergarten (ECLS-K:2011) http://nces.ed.gov/ecls/	Assessment of how childhood, parent, school, and community factors affect childhood development, early learning, and school progress	Students, parents, and teachers of a nationally representative cohort of elementary school students in kindergarten in 2011	Cohort surveyed in kindergarten (2010-2011), 1st grade (2011-2012), 2nd grade (2013), 3rd grade (2014), 4th grade (2015), and 5th grade (2016)	Data tables available; occasional analysis of elementary school instruction, or trends in racial and ethnic groups (e.g., 2010, 2007, 2006)	2012	2013
High School Longitudinal Study (HSLS:09)	When and how students decide on secondary courses, choose among postsecondary options, and consider which career(s) to pursue	Cohort of 9th grade students, their parents, math and science teachers, school counselors, and school administrators	Cohort surveyed in 2009 (9th grade), 2012 (11th grade), 2013/2014 (end of secondary education) and 2015 (2 years after secondary school)	Data tables available; too early for in-depth analysis	2012 collection (in 11th grade)	2013/2014 (end of secondary education)
National Assessment of Educational Progress (NAEP)–Background Questionnaires in Mathematics and Science http://nces.ed.gov/nationsreportcard/	Background questionnaires for students, teachers, and schools to provide contextual information that explains NAEP achievement data	Students, teachers (including special and bilingual education), and principals in grades 4, 8, and 12 in schools of a nationally representative sample	Every 2 years for mathematics, every 4 years for science	*The Nation's Report Card,* various studies of achievement gaps, charter and private schools, students with disabilities, and English language learners	2011 for mathematics and science	2013 for mathematics, 2015 for science

Survey and URL	Description/ Focus	Respondents	Frequency of Administration	How Data Are Compiled	Most Recent Data Collection	Next Expected Data Collection
National Assessment of Educational Progress (NAEP)–Technology and Engineering Literacy http://nces.ed.gov/nationsreportcard/	Continuing assessment of what America's students know and can do in engineering and technology	Students in grades 4, 8, and 12 in schools of a nationally representative sample	Every 2 years	*The Nation's Report Card,* various studies of achievement gaps, charter and private schools, students with disabilities, and English language learners	Under development	2014
National Education Longitudinal Study (ELS:2002) http://nces.ed.gov/surveys/els2002/	Assessment of how achievement, interest, and aspirations in high school affect outcomes in higher education and in the workforce	Students, parents, teachers, librarians, and schools of a nationally representative cohort of high school students in 10th grade in 2002	Cohort surveyed in 2002 (10th grade), in 2004 (12th grade), in 2006 (2nd year out of high school), and 2012 (8th year out of high school)	Data tables available; occasional analysis of high school seniors and those who drop out (e.g., 2009, 2008)	2012	Unknown
National Household Education Surveys (NHES) http://nces.ed.gov/nhes/	Early childhood development, school choice decisions, parent and family involvement, and postsecondary planning activities of school-aged children and their families	A survey of a representative sample of households across the country	Every 2 years (no 2009 data available as the survey was designed)	Regular use in analysis of adult education, before- and after-school activities, and family involvement in education	2012	2014

Survey and URL	Description/ Focus	Respondents	Frequency of Administration	How Data Are Compiled	Most Recent Data Collection	Next Expected Data Collection
Schools and Staffing Survey (SASS) http://nces.ed.gov/surveys/sass/	Four interrelated surveys: school, teacher, principal, and school district questionnaires provide descriptive data on the context of elementary and secondary education. Topics include teacher demand, teacher and principal characteristics, general conditions in schools, and teacher compensation	A survey of a representative sample of schools across the country	Every 4 years	Regular use in analysis of school safety, and school libraries	2011-2012	2015-2016
Schools and Staffing Survey–Principal Follow-up Survey (PFS) http://nces.ed.gov/surveys/sass/	Principal retention and mobility	Survey sent to the principal of every school selected for the previous year's SASS	Every 4 years, in the year following the SASS	Data tables available; occasional use in analysis of principal attrition	2008-2009	2012-2013
Schools and Staffing Survey–Teacher Follow-up Survey (TFS) http://nces.ed.gov/surveys/sass/	Teacher retention, teaching status and assignments, and information on decisions to change schools	Survey sent to the principal of every school selected for the previous year's SASS	Every 4 years	Regular use in analysis of teacher attrition and teacher qualifications	2008-2009	2012-2013

ACKNOWLEDGMENTS

This report has been made possible by the important contributions of many individuals and organizations. First, we acknowledge the sponsorship of the National Science Foundation, and especially the steadfast support of Janice Earle, program director in the Education and Human Resources Directorate.

The committee also thanks the National Research Council (NRC) staff of the Board on Science Education and the Board on Testing and Assessment who worked on this project: board directors Martin Storksdieck and Stuart Elliott for providing valuable oversight and general wisdom during the committee deliberations and report writing; senior program officer Natalie Nielsen, who served as the study director, for her superb accomplishments and tireless efforts in taking the committee's ruminations and transforming them into coherent prose; and program associate Rebecca Krone for her unparalleled organizing skills and consummate professionalism. We also thank Mac Hird for providing research assistance during his summer internship, and NRC staff Eugenia Grohman for editing the report, Kirsten Sampson Snyder for shepherding it through the review process, and Yvonne Wise for facilitating its production.

This report has been reviewed in draft form by individuals chosen for their diverse perspectives and technical expertise, in accordance with procedures approved by the NRC's Report Review Committee. The purpose of this independent review is to provide candid and critical comments that will assist the institution in making its published report as sound as possible and to ensure that the report meets institutional standards for objectivity, evidence, and responsiveness to the study charge. The review comments and draft manuscript remain confidential to protect the integrity of the deliberative process.

We thank the following individuals for their review of this report: Elaine Allensworth, Chicago Consortium of School Research, The University of Chicago; Deborah Loewenberg Ball, School of Education and Institute for Social Research, University of Michigan; William B. Bridges, School of Engineering, Emeritus, California Institute of Technology; Gail Burd, Department of Molecular and Cellular Biology, University of Arizona; Margaret E. Goertz, Graduate School of Education, University of Pennsylvania; Joan Herman, Center for Research on Evaluation, Standards, and Student Testing, University of California, Los Angeles; John G. Hildebrand, Department of Neuroscience, School of Mind, Brain, and Behavior, University of Arizona; Sharon Lohr, Survey Methods Unit, Westat, Rockville, Maryland; Jonathan Osborne, Graduate School of Education, Stanford University; Steven A. Schneider, Science, Technology, Engineering, and Mathematics, WestEd, Redwood City, California; Jerry D. Valadez, Central Valley Science Project, California State University, Fresno; Matthias Von Davier, Research and Development Division, Educational Testing Service, Princeton, New Jersey; Iris R. Weiss, Office of President, Horizon Research, Inc., Chapel Hill, North Carolina.

Although the reviewers listed have provided many constructive comments and suggestions, they were not asked to endorse the conclusions or recommendations nor did they see the final draft of the report before its release. The review of this report was overseen by Stephen E. Fienberg, Department of Statistics, Carnegie Mellon University, and Lauress L. Wise, Education Policy Impact Center, HumRRO, Monterey, California. Appointed by the NRC, they were responsible for making certain that an independent examination of this report was carried out in accordance with institutional procedures and that all review comments were carefully considered. Responsibility for the final content of this report rests entirely with the authoring committee and the institution.

Adam Gamoran, *Chair*
Committee on the Evaluation Framework for
Successful K-12 STEM Education